Lepidoptera - Advances in Ecology, Conservation, and Taxonomy

Edited by Alessio Vovlas

Published in London, United Kingdom

Lepidoptera - Advances in Ecology, Conservation, and Taxonomy
http://dx.doi.org/10.5772/intechopen.1008095
Edited by Alessio Vovlas

Contributors
Ali Rajabpour, Aquib Majeed, Avisek Patra, Ayesha Tariq, Chen Hang, Ding Weifeng, Farrukh Baig, Fatemeh Yarahmadi, Hafsah Ghaffar, Han Shichang, Ling Xiaofei, Lu Qin, Mirza Abdul Qayyum, Muhammad Asif Farooq, Sajad Hussain Parey, Tabraq Ali, Unsar Naeem-Ullah, Wang Weiwei, Zahid Hussian, Zhang Jinwen

First published in London, United Kingdom, 2025 by IntechOpen
IntechOpen is the global imprint of INTECHOPEN LIMITED, registered in England and Wales, registration number: 11086078, 167-169 Great Portland Street, London, W1W 5PF, United Kingdom

For EU product safety concerns: IN TECH d.o.o., Prolaz Marije Krucifikse Kozulić 3, 51000 Rijeka, Croatia, info@intechopen.com or visit our website at intechopen.com.

British Library Cataloguing-in-Publication Data
A catalogue record for this book is available from the British Library

Lepidoptera - Advances in Ecology, Conservation, and Taxonomy
Edited by Alessio Vovlas
p. cm.
Print ISBN 978-1-83635-611-0
Online ISBN 978-1-83635-610-3
eBook (PDF) ISBN 978-1-83635-612-7

If disposing of this product, please recycle the paper responsibly.

Meet the editor

Alessio Vovlas received a Ph.D. in Science and Technology from Turin University, Italy, in 2014 with a thesis on "Evolutionary Biology and Biodiversity Conservation", and an MS in Natural Science from Bari University, Italy. He is an effective member of A. P. S. Polyxena, an NGO that is part of the Butterfly Conservation Europe, European Citizen Science Association (ECSA), and Societas Europaea Lepidopterologica (SEL). He is also a member of the Society for Conservation Biology and part of the membership committee of the International Union for Conservation of Nature (IUCN) Commission on Education and Communication (CEC). Dr. Vovlas' primary research interests are molecular biology, ecology, zoology, science communication, and education. He has published several scientific, national, and international publications and coordinated some nature conservation and citizen science projects in natural park areas. Since 2017, he has been a member of the Education and Communication Commission (CEC) of the International Union for Conservation of Nature (IUCN). He is also currently involved in a European Food Safety Authority (EFSA) project aimed at developing investigation tools for pests of vegetable and fruit plants.

Contents

Preface

The order *Lepidoptera*, encompassing both butterflies and moths, represents one of the most diverse and evolutionarily successful lineages of insects on Earth. Beyond their aesthetic appeal and cultural symbolism, *Lepidoptera* play a pivotal role in global ecosystems as pollinators, herbivores, and bioindicators. Their morphological diversity, behavioral complexity, and ecological sensitivity render them an exceptional model for studying fundamental biological processes, from development and adaptation to speciation and conservation dynamics. This volume, *Lepidoptera – Advances in Ecology, Conservation, and Taxonomy*, seeks to advance the scientific discourse surrounding these remarkable organisms through an integrative exploration of their biology at molecular, structural, and ecological levels.

Recent years have witnessed a profound transformation in lepidopteran research, driven by innovations in molecular biology, genomics, and imaging technologies. The intricate architecture of butterfly and moth wings, long admired for their colors and patterns, has emerged as a powerful field of molecular and biophysical inquiry. Studies on wing microstructures, particularly at the nano- and micro-scale, have revealed how structural coloration and functional adaptations arise from finely tuned interactions between chitin, pigments, and light. The chapters within this book delve into these molecular aspects, highlighting the interplay between genetic regulation, cellular processes, and the emergent physical properties of wing scales. Such research not only enhances our understanding of *Lepidoptera* themselves but also informs biomimetic design, photonics, and materials science.

Equally significant is the examination of the microstructural dynamics involved in eclosion and wing expansion, processes that mark the culmination of metamorphosis and the transition to adult life. The physiological mechanisms that govern wing inflation, cuticular hardening, and scale differentiation are explored here as critical determinants of flight capability and survival. By analyzing these processes at multiple biological scales, the volume underscores the delicate balance between developmental precision and environmental sensitivity that defines the life history of *Lepidoptera.*

In parallel with these molecular and developmental perspectives, the book emphasizes the ecological and conservation dimensions of lepidopteran research. *Lepidoptera* have long served as key indicators of environmental change, their population dynamics reflecting shifts in climate, land use, and habitat integrity. Comprehensive monitoring programs, supported by both traditional field observation and emerging digital and genetic tools, are increasingly vital for assessing biodiversity and ecosystem health. Through a synthesis of ecological modeling, citizen science, and molecular taxonomy, this volume illustrates how integrative methodologies can enhance both species identification and conservation strategies.

Taxonomy remains the foundation upon which all other studies of *Lepidoptera* are built. As molecular data refine our understanding of phylogenetic relationships, new insights continue to reshape long-standing classifications and reveal cryptic diversity within familiar taxa. The chapters devoted to systematic and taxonomic advances in this collection aim to bridge classical morphological approaches with genomic and phylogenomic frameworks, promoting a cohesive and comprehensive vision of Lepidopteran diversity.

Ultimately, *Lepidoptera – Advances in Ecology, Conservation, and Taxonomy* aspires to serve not merely as a compilation of contemporary research but as a platform for interdisciplinary dialogue among molecular biologists, ecologists, taxonomists, and conservation practitioners. By integrating multiple dimensions of inquiry, from the molecular structure of wing scales to the global patterns of species distribution, this book underscores the necessity of a holistic approach to understanding and protecting *Lepidoptera* in the Anthropocene.

In presenting these advances, we also reaffirm the enduring significance of butterflies and moths as both scientific subjects and symbols of transformation, resilience, and ecological harmony. Their fragile beauty and complex biology remind us that the preservation of biodiversity is not only a matter of scientific responsibility but also a matter of cultural and ethical stewardship toward the natural world.

Alessio Vovlas
Italian National Research Council,
Bari, Italy

Chapter 1

Fading Colours: Human and Natural Influences on the Decline of Butterfly Diversity

Unsar Naeem-Ullah, Ayesha Tariq, Mirza Abdul Qayyum, Muhammad Asif Farooq, Farrukh Baig and Hafsah Ghaffar

Abstract

Butterflies (Lepidoptera: Insecta) are well-known as pretty, delicate insects found on every continent except Antarctica. With approximately 20,000 described species, many more remain undiscovered. However, the industrial revolution's onset has led to significant human interference, altering ecosystems worldwide. Consequently, butterflies face survival challenges, resulting in a rapid decline in their population and diversity. Human activities continue to degrade the environment, worsening the loss of butterfly species. If left unchecked, future generations may only glimpse these beautiful insects in pictures. Butterfly declines have been rampant in recent decades and have led to concerns regarding the area of interest of scientists and other conservationists. This chapter shows how anthropogenic [human-mediated] pressures, of which the clearance of forests, agricultural exploitation with pesticides and fertilizers, urbanization, and road routes, as well as the impacts of global climatic change, are removing the habitats of butterflies and interposing on their food resources. The present review also evaluates environmental ones such as storm incidents and phenological and seasonal changes in plants, which may undermine butterfly survival and reproduction. These forces tend to act synergistically, and hence the observed decline becomes intense. There is also, in the chapter, a focus on the ecological interactions that are complex and connect butterflies to many other groups of insects, birds, and plants, and point to the idea that the loss of a butterfly population can have echoes throughout these interacting communities. In sum, these considerations light up the dual functions of human activity and ecological changes in the continuous diminishing of butterfly diversity of global ecosystems.

Keywords: anthropogenic, climate change, Lepidoptera, natural disasters, Rhopalocera

1. Introduction

Out of the insect order Lepidoptera, there exists an order of Rhopalocera that forms the main group of butterflies [1]. This taxon is consistently endorsed in its wing pigmentation color and the intricate pattern of the wing that has both an esthetic and practical role in thermoregulation of the animal, camouflage, mate selection, and avoidance by predators. More than their appearance, butterflies are involved in important ecological functions, which include herbivores, pollinators, and prey, and thus are essential in the stability and resistance of the terrestrial ecosystem [2].

Suborder Rhopalocera is further divided into six major families, namely Papilionidae, Pieridae, Lycaenidae, Nymphalidae, Hesperiidae, and Riodinidae, based on their own morphological, behavioral, and ecological adaptations [3]. Worldwide, there is a very extensive distribution of butterflies except in high latitudes such as the poles, with the highest biodiversity occurring in tropical regions of China, central Africa, and South Asia. Extraordinary biodiversity hotspots are discovered in countries such as Pakistan, India, and Indonesia, implying that the countries need to be regarded as important in the preservation of global butterflies [4].

Butterflies have been facing a rate of decline that has never been seen before, and due to their role both ecologically and their importance culturally, this is alarming. Numerous anthropogenic forces, such as habitat loss and fragmentation, rapid urbanization, the intensification of agricultural activities, and climatic changes, as well as pollution and the distribution of invasive species, are causing a multidimensional problem threatening their survival [5].

These drivers often combine their effects and dwarf the effects of other individual stressors. Considering an example, habitat fragmentation not only diminishes the quantity of the habitats but also isolates the populations and aggravates the occurrence of an effect of the edge and promotes the infiltration of new, exotic species that generally disrupt the local communities. Besides, climate change modifies temperature regimes, precipitation patterns, and synchrony of the phenology of butterflies and their host plants, changes species distributions, and thus complicates conservation practices.

Therefore, whenever these threats have been identified individually by many studies, a great deal is still lacking in terms of how they interactively affect each other and their evolutionary implications in the long run. There is a need to fill this gap in order to come up with complete and effective conservation approaches that should protect not only butterfly species richness but also the functional diversity and the ecological interactions that contribute to the health of the ecosystem. Greater appreciation is thus required on an integrative level of these dynamics to ensure that additional losses of the diversity of butterflies are avoided.

2. Butterfly diversity in the world

Butterfly diversity is nonuniform, with about 12,119 species or more currently estimated to be found. In small areas and tropical and subtropical high-altitude areas in different parts of the world that are considered to be hotspots of butterfly biodiversity because of their richness, rareness, and phylogenetic distinctiveness [6]. Major areas are as follows: the Hengduan Mountains in China, which are a hotspot of Papilionidae diversity [7]; the Comoro Islands, which have endemic species; and China, where almost 10% of all butterfly species may be found [8].

High species turnover and richness are also seen in tropical mountains around the world [9]. Such regions mentioned in literature include the Kumaon region in India, which indicates butterflies as a bioindicator [10], and Arunachal Pradesh, where 281 species are reported [2]. There have been studies across Europe in Portugal [11] and Belgium [12] to guide conservation planning, as in Asia, where butterfly species diversity and rarity have been recorded in Japan's Aokigahara woodland [13].

3. Role of butterfly in natural ecosystem

The butterflies are important in their ecosystems because of the provision of pollination and biodiversity, as well as the general functioning of an ecosystem.

In addition to bees, there is also the important role of butterflies in the reproduction of plants, especially in agroforestry and in some plant species [14, 15]. They add value to the ecosystem services and could be used as bioindicators of the health of the pollination [16]. Elevated biodiversity of butterflies is indicative of complex and stable ecosystems, and measures that encourage butterfly populations, such as maintaining semi-natural habitats or restoring prairie strips, can increase butterfly numbers [17, 18]. Butterflies, as pest and pollinators, exist in the food web and are fundamental to the accessibility and the quality of the resources [19].

4. Decline and reasons of decline

Butterflies are facing population decline, which endangers biodiversity and ecosystem integrity in several parts of the world as a result of an intricate assemblage of anthropogenic and natural processes [20]. With respect to anthropogenic factors, the most common ones, shrinking and fragmentation due to urban sprawl and land use, are the main causes leading to decreasing connectivity between habitats and growing local extinctions [21]. Promoters of agricultural intensification, especially pesticide application and nitrogen deposition [22], contribute further to imbalances in butterfly populations, also reducing larval survival and worsening the quality of the host plants [23]. One vivid case is the case of monarch butterflies, which began to decline in North America mostly because of the reduction of milkweed after the introduction of herbicide-resistant crop production [24].

These pressures are aggravated by the natural forces, particularly climate change, due to the modification of rainfall patterns, generation of improper microclimates, and disturbance of butterfly-to-host plant phenological synchrony [25]. Other adversities are intrusive species, new diseases, and alterations in migration and reproductive trends, including diminished monarch migration [26]. It is these threats that tend to work collectively and not independently of each other and increase their compiled full effect.

These are some of the major factors bringing about declines in butterflies, which are analyzed more in the following sections.

4.1 Anthropogenic

Ecosystem depends on insect biodiversity, especially pollinators, and butterflies are among them. Sadly, human actions such as urbanization, pollution, deforestation, pesticide usage, agricultural infestations, and climate change are progressively endangering it [27]. Many human actions, including pollution, invasive species, land-use changes, and climate change, are gradually under threat of insect biodiversity. Ref. [28] reports that the world population is growing and projected to increase for another 5060 years, which results in major environmental changes. Major stressors on biodiversity [29] are well-documented from the interaction of agriculture, grazing, industrial development, vehicle traffic, infrastructure expansion, and noise pollution.

4.1.1 Deforestation

Deforestation causes major disturbance to forest ecosystems, especially silvicultural habitats, as it will remove shade, humidity, and certain host plants that are essential to many of the butterflies that depend on forests. Further, while exposed or light-dependent species may take advantage of some reduced forest coverage from the opening created by habitat clearing, the long-term consequences of deforestation are serious. Deforestation can lead to fragmented and isolated populations, increased wind exposure, increased stress from temperature, increased exposure to predators, and pressures that all negatively affect butterfly survival.

The high biodiversity of tropical forests also supports coevolutionary relationships of butterflies with dark organisms that include predators, mutualists, potential mates, and competitors. This leads to not only the quantity of species but also extensive variation in species' phenotypic and behavioral characteristics involved with coloration strategies [30]. Deforestation in tropical forests resulting from agricultural expansion is moving rapidly and is causing profound changes to local biodiversity [31]. Additionally, studies have shown that the human-caused interference contributes to the selective pressure on butterfly populations and declining diversity of coloration and defense mechanisms [32]. These changes highlight the complexity of habitat loss on species' richness on both a remarkable scale and level of ecological complexity.

The case of the endangered Kaiser-i-Hind butterfly [*Teinopalpus imperialis*] is a good example of this tendency; its populations have dramatically dropped over its entire range in the Eastern Himalayas and Southeast Asia due to massive deforestation [33]. These trends are indicators of the complexity and far-reaching implications of deforestation on butterfly diversity and the need for holistic, all-embracing conservation efforts.

4.1.2 Alteration of ecosystems and natural landscapes [urbanization, urban lights]

Urbanization, infrastructure expansion, and rising demand for goods and resources all speed up the rate of land change. These pressures put progressively more strain on natural ecosystems all across [34]. Human-induced changes sometimes start complicated, cascaded effects with both short- and long-term consequences.

Of course, that is alarming because insects provide vital ecosystem services and are important to other ecosystem services, including pollination, decomposition, and as base consumers [35]. There are studies that suggest the effects of landscape change are severe. Butterfly populations, which are identified as at-risk, are declining by 8% per year, but over a 20-year time span, this equates to an 83% decline in total abundance. The rate of decline for specific at-risk butterflies is substantially higher than the global average decline of 1–3% per year for butterfly communities [36, 37].

For instance, great road networks span developed nations worldwide. New research on monarch butterflies [*Danaus plexippus*], though, has found that fewer than predicted numbers of eggs and larvae exist in roadside environments, therefore implying that female monarchs avoid egg laying near roads or possibly are killed by automobiles [38].

4.1.2.1 Urbanization

Butterflies, among other insects, live in a great range of environments and are becoming more and more vulnerable to several stresses. Among the most serious threats are habitat fragmentation, degradation, and direct loss [39].

DOI: http://dx.doi.org/10.5772/intechopen.1011999

For roadways, structures, and agricultural use, urban growth and infrastructure improvements keep eating natural environments, therefore drastically affecting insect populations.

Although some insect species may live in small patches of nature, these separated groups are especially susceptible to chance environmental occurrences and loss of genetic diversity. Bigger populations, in contrast, are more resilient and flexible. Urban development is regrettably quickly depleting natural regions all around. In some areas, the condition is quite bad. With another 19 classified as threatened, 20 butterfly species are already extinct in Flanders [Belgium], among the most badly damaged areas in Europe [40]. Several butterfly species have little habitats and are very vulnerable to local environmental disruption. Therefore, they provide significant indications of ecosystem health [41]. The detrimental impacts of urbanization on butterflies all throughout are also verified by the study [42], which also emphasizes the degree of its impacts on this group of bioindicators.

This highlights the fragmentation effects of urbanization itself, i.e., urbanization also results in faster local extinction, in particular of specialist species. As an example, urbanization and destruction of habitat have been strongly implicated in the extinction of the Violet Copper [*Lycaena helle*] in some parts of Western Europe. Combined, these effects raise a need to find land-use planning that will explicitly accommodate the conservation of butterflies to retain existing populations.

4.1.2.2 Urban lighting

Artificial Light at Night [ALAN] is dramatically altering natural light cycles globally, with some of the potential consequences being large changes in insect behavior and survivorship [43]. Among the most dramatic and ecologically significant effects are the disruptions of the flight-to-light behavior and the changes that can lead to many insects getting lost.

All butterflies use daylight [most are diurnal], but some rely on the night period for critical processes such as pupation [transforming from a caterpillar to a butterfly], orienting themselves, and roosting in relative safety. Many urban and suburban areas have excessive lights that hinder or eliminate the ability of insects to complete their regular cycles, with many observations recording that these effects led to confusion and disorientation, developmental delays, and higher mortality rates [44]. In the tropics, we will also often find warm microclimates that influence the community composition of butterflies, influenced by habitat type.

For example, the savanna woodlands have higher butterfly diversity when compared to the nearby gallery forests or riparian areas. The reality of how urban microclimates and light pollution structure populations is illustrated in [45].

Overall, the studies show that ALAN is an invisible, but real, stressor and acts to further limit both population growth and resilience in already fragile butterfly populations. For example, urban light pollution means declines in the *Zerynthia polyxena* [Southern festoon] will be solely from their reliance on undisturbed nighttime environments to successfully pupate. Managing light, therefore, should be part of planning for urban environments to limit reductions in butterfly populations.

4.1.3 Pollution

Environmental pollution is currently heralded as one of the main causes of biodiversity loss on a global scale. This includes drastic declines in insect diversity and abundance [46]. Here we discuss examples of pollution that are most deleterious to butterflies, i.e., pesticides, fertilizers, air pollutants, noise, and light pollution [47].

Particulate matter [PM] is concerning to butterflies, as it is an air pollutant that has direct health impacts on both adults and larvae. PM exposure has been shown to directly damage DNA in larvae, which allows PM to contribute to local declines, possibly even extinction [48]. For example, studies on common widespread species suggest that genotoxic effects of PM could explain reductions in populations in otherwise common and healthy butterflies.

4.1.3.1 Noise pollution

Butterflies do not utilize acoustic communication modes, but noise pollution is also likely to reduce their fitness by compromising their orientation and/or ability to find mates or escape from predators. Constant high roadside noise, an overlooked consideration in biodiversity conservation, increases stress in larvae, e.g., monarch butterfly larvae stressed significantly more in response to traffic noise; this is now considered a major consideration in larval rearing and survival [49, 50]. Noise pollution may even affect butterflies indirectly *via* stressed plants and reductions in plant health and distribution, creating further habitat loss.

4.1.3.2 Air pollution

Pollution [including smog and heavy metals] presents an obvious risk to butterfly populations [51]. Butterflies are a sensitive bioindicator, and smog and other toxic particles impair their survival, growth, and reproductive output [52]. In addition, pollutants such as lead are linked to impairments in cognitive functions such as navigation or finding host plants, as demonstrated in a variety of insect studies [53].

Overall, these findings suggest pollution acts as a largely underappreciated but harmful threat to butterfly persistence. For example, urban air pollution has been suggested as a possible cause of the decline of the Common Blue [*Polyommatus icarus*] in several European cities. These impacts stress the need for a more integrated approach to pollution reduction in butterfly conservation work.

4.1.4 Agriculture infestation

Agricultural habitats are generally recognized to be among the rapidly declining ecosystems [54]. These declines are primarily attributed to intensive agricultural systems, habitat conversion, and land-use conversion; however, such agricultural systems are also associated with habitat fragmentation, pesticide practice, and monoculture farming systems. Not only does sustainable agricultural habitat pose a significant biodiversity threat [55]. It also continues to destroy invertebrate diversity to the point that ecologically important ecosystem functions that additionally rely on invertebrates, like pollination and pest management, are declining even further. Intensification of agriculture is an acknowledged primary long-term threat to butterflies in Europe. After agricultural intensification, butterfly species richness and abundance are lost to habitat loss, use of pesticides, and associated by-products of monoculture farming [56].

There have been significant declines in insects across central, western, and northern Europe. Currently, this decline is presumed to be highly attributed to agricultural intensification [23].

4.1.4.1 Habitat destruction

Habitat loss and degradation are widely recognized as the primary threat to global biodiversity [57]. For butterflies, habitat loss resulting from urbanization, agricultural intensification, and chemical use [e.g., pesticides], and climate change is one of the top reasons for their decline [58]. Any destruction or serious degradation of meadows, grasslands, bogs, and forest edges will directly remove the breeding, feeding, and overwintering resources they rely upon. Because butterflies are sensitive indicators of ecosystem health and also important pollinators and herbivores, they deserve conservation attention. Unfortunately, they are one of the most threatened orders of insects. In Great Britain, for instance, butterflies are vanishing more than many of its taxa [59].

Habitat fragmentation enhances these losses, as habitat patches shrink in size and transitional zones, as vegetative elements, lose circulation and dispersal [60]. Designation of protected areas is not sufficient to conserve butterflies; active management of habitat is essential to conserve butterfly diversity [61]. Specialist butterflies, which are dependent on specific habitats and microclimates, are at highest risk. The specialist species tend to exist in small, separated populations, with few disappearing host plants and less microclimate stability, which have negative consequences on their potential breeding and migration [62]. Also, fragmentation has a negative effect on movement, which reduces the genetic diversity of populations and their ability to adapt to changing environments; consequently, both of these induce a heightened chance of extinction risk.

These trends illustrate that habitat destruction remains a silent but significant threat to biodiversity for butterflies on an overall basis, especially at local scales. The most glaring example of how sensitive specialists can disappear is the drastic reduction of the Marsh Fritillary [*Euphydryas aurinia*] population across Europe, which epitomizes the fact that specialists simply do not exist when habitat connectivity is lost. Proactive measures for habitat restoration and connectivity of habitats are required urgently to avoid ongoing losses.

4.1.4.2 Pesticide use

Another leading cause of butterfly declines and reductions in diversity at a global scale is pesticide use. Although habitat loss and climate change are fundamental threats, pesticides, particularly insecticides and herbicides, have far more direct and often immediate consequences on butterfly survival [63]. Insecticides can cause both direct mortality and sub-lethal effects, including reduced fecundity, lifespan, reduced foraging, and impaired reproduction [64]. Neonicotinoids, notoriously harmful to bees, disrupt similar behaviors in butterflies that are fundamental to the population's survival.

Herbicides indirectly impact butterflies by eliminating nectar [and larval] host plants [65]. A prime example is the decline of milkweed due to herbicide application in North America and the concomitant decline of the monarch butterfly [*Danaus plexippus*]. In addition, an added risk is presented through herbicide or insecticide spray drift, as downwind monarch eggs, larvae, and adults are exposed to a lethal dose of either application [66].

Overall, it is clear that pesticides are acting as a hidden but undeniable and substantial driver of butterfly declines, acting both through direct toxicity and ecosystem simplification. Thus, the significant decline in monarchs is a glaring reminder of their dependency upon undisturbed host plant networks. The implementation of further regulatory action on the uses of pesticides and better pest management strategies can be an important step toward conserving butterfly taxa.

4.1.4.3 Monoculture

Monoculture is a principle that is often practiced in agriculture, which harmfully impacts butterfly populations and biodiversity [67]. Monoculture features the production of a single crop over a large area. Monoculture leads to habitat loss and decreases the availability of food sources, lessening ecological balance [68]. Habitat loss directly lowers the living spaces and food resources for butterflies [69]. Many butterflies depend on specific host plants, which they use for their larvae, or plants that are sources of nectar for adult butterflies. Several studies have found evidence that extensive deforestation resulting in agricultural monocultures, such as oil palm plantations, may destroy habitats and lead to reduced biodiversity. An example is agriculture with agroforestry orchards, which would strengthen greater diversity in tropical butterflies compared to monoculture plantations or other systems [67]. Butterflies often rely upon a variety of plants during various life stages [20], but monoculture farming provides limited food source availability, often less food and resources for larvae and adults [70].

Monocultures of one crop limit the nectar resources for adult butterflies and eliminate the host plants necessary for butterfly larvae, leading to lower butterfly abundance and diversity. Forest and grassland habitats are expected to provide higher species richness and abundance than agricultural habitats, as they have more diversity and more stable conditions [71]. In agriculture, it is possible to increase butterfly diversity in agroforestry systems with the integration of additional habitats and food sources. Agroforestry orchards, for instance, show more butterfly diversity than monocultures [72].

4.1.5 Technological hindrance

There is evidence indicating that mobile phone radiation has harmful impacts on various types of organisms. For instance, mobile phone electromagnetic waves [EMR] severely disrupt the behavior of honeybees, who exhibit typical worker piping behaviors [73]. This work studied bees but raises the question of whether butterflies and other insects may respond similarly.

In addition to interference with behaviors, EMR can affect morphology and biological functions, like reproduction, in insects [74]. There's a study that showed that EMR has a negative impact on insect distributions. Some examples include the brown-veined white butterfly *Belenois aurota* [Fabricius, 1793] and darkling beetle *Nesotes azoricus* [Crotch, 1867], which are both critically endangered along with numerous other insect species and are found only within areas deficient in EMR in the Hook Bridge area [75]. Although there is no direct link between mobile phone adaptations in the sense of evolutionary adaptations of mobile phone usage necessarily causing butterfly declines.

5. Natural ecosystem

The natural ecosystem cycle considerably affects butterfly populations through several interconnected factors such as habitat availability, host plant dynamics, and interactions with other species [76].

5.1 Climate change

The occurrence of climate change influences the numbers of butterflies by changing their distribution patterns, access to habitat, or phenology [77]. The research indicated by [78] showed that the richness of butterfly species and the ability to fight off extinctions due to warmer weather decrease as the temperature rises. Nevertheless, such responses are highly differentiated; whereas one species adapts or changes its distribution, others will decline because of poor dispersal potential or habitat specialization [79]. In the Picos de Europa mountain range, the overall abundance of butterflies has gone down by 45% over 9 years, of which 26% of the species had been significantly affected by the joint impacts of climate change and loss of habitat [80]. Also, the famous monarch butterfly [*Danaus plexippus*] has suffered sharp declines, in part due to changes in climate variation in the range of overwintering locations in Mexico [63]. Moreover, data from the Netherlands support these results, where egg-larvae hibernators decreased by 63% in 13 years, whereas there was no significant trend in adult-pupa hibernators [81].

These findings demonstrate that climate change poses both direct and indirect threats, which might challenge the adaptability of the butterflies at various stages of life. The noticeable loss of overwintering egg-larvae of prey species, such as the small tortoiseshell [*Aglais urticae*], is a reminder of the power of very localized climatic impact in driving even abundant species to extirpation. These species-specific vulnerabilities have to be taken into consideration in adaptive conservation measures.

5.2 Floods and storms

We are now coming to realize that climate change [82] not only raises the global temperatures but also makes floods and hurricanes more common and severe [83]. Such extreme weather conditions directly attack the habitat of butterflies, and the resultant change can alter their entire life cycle [84]. Take the example of breeding spots of a butterfly, such as grassland or forests; in case such areas flood or are struck down by a storm, they may collapse. In one study of hedgerows, the researcher discovered that the importance of hedgerows as a hotspot for wildlife depends on their construction. When an extreme storm splits those hedgerows into pieces, the opportunities for butterflies to dwell in them decrease, and the same happens to butterfly populations [85].

Floods cause damage of their own. Eggs and larvae are washed away when waters increase, and this wends development. Host plants may be knocked over by storms, larvae may be food-deprived, and butterfly reproduction may decline [86]. With each subsequent rise of the water, butterfly communities move to a new place, and some of them do not recover completely after losing their habitats, becoming at-risk of local extinction. Even alterations in land management add on. Managed grasslands, such as ski slopes, wither and go dry when left to fester, affecting butterfly diversity. This process can be fast-tracked by post-flood and storm cleanup activities, which leave even more open space and less food available to adult butterflies [87].

All this emphasizes one important fact: the floods and storms caused by climate change destroy the habitats of the butterflies directly and deteriorate them indirectly, and, therefore, increase the vulnerability of the butterfly populations.

5.3 Competition with invasive species

Invasive species are one of the top threats affecting global butterfly biodiversity [88] and ecosystems primarily because their impacts occur on multiple levels. They modify habitat structure, consume resources [often in aggregate with the facilitation of other invasive species] from limited resources, and modify behaviors and life-history traits in native insects. As a result, it is imperative for conservation programs to navigate the complex feedback loops associated with invasive populations and native insect communities. [81]. Consider invasive plants that can drastically alter habitat composition and structure while occupying and degrading native butterfly habitats. An example is *Lantana camara*, an aggressive weed found throughout tropical understories that devours access to essential host plants and microhabitats. Such invasive taxa alter top-down impact *via* competitive exclusion and have downward pressure *via* habitat alteration [89].

These examples illustrate not only how invasive species are silently changing butterfly habitats but also how they can undermine even well-intentioned restoration projects by changing a habitat before restoration can happen. Fender's blue [*Icaricia icarioides fenderi*] decline in Oregon prairies shows how invasive grasses can hoist rare specialists. It is critical that conservation efforts tightly couple invasive species control and habitat management, effectively and consistently, to support butterfly populations that are in flux.

5.4 Volcanic eruptions

The ecosystems seriously affected by volcanic eruptions and butterfly populations in particular can be altered by habitat change and atmospheric influence [90]. Although no direct studies of the effects of volcanic eruptions on biodiversity exist, based on the similarity with the most relevant studies, one can assume that all these factors have the same effect on butterfly populations [91].

The volcanic eruptions have the potential to destroy the landscapes, cover them with ash, and change the vegetation structure, which is also essential to the butterfly habitats [20]. Their effects depend on tephra type, such as crystallinity or chemical composition, because vegetative structure changes as a result of deposition of the ash can influence butterfly diversity. Research indicates that the dynamics of the butterfly population are associated with encroachment on vegetation [92].

Butterfly declines are possible as a result of volcanic eruptions due to habitat change and climate change. Although the immediate consequences of volcanic eruptions on butterfly populations are under-researched, the insight into the mechanics is the information that can be used toward conservation.

5.5 Change in Earth's magnetic field

There are possibilities that the Earth's magnetic field and its variations can affect the way animals navigate, especially the migratory ones. Among these species is the monarch butterfly, which is unique because of its long-distance migrations. Alterations in the magnetic field of the Earth [93], either on a natural basis or due to anthropogenic changes in electromagnetic fields [EMF], technically have the possibility of influencing populations of monarch butterflies by either interfering with their navigation skills or otherwise due to their physiology [94].

A difference in Earth's magnetic field might alter the effectiveness of the internal compass of the monarch and disorient them, as they might fail to reach their destination in overwintering [95]. Although one study pointed out that a change in the geomagnetic field had minimal impact on the overwintering range of Eastern North

American fall migratory species of monarch butterflies, implying that they had no magnetic map sense used to navigate.

EMF can affect cell processes involving calcium channels and signaling, metabolism, etc. These alterations may impact other biological aspects of butterflies, such as the specimen's development, reproduction, and behavior [96]. The changes in the habitat, caused by the use of herbicides and deforestation, are major factors that can threaten the population of monarchs [63]. These can combine with any effects that the magnetic field shift causes, making the situation even more complicated in regard to butterflies [24].

5.6 Forest fires

Butterfly populations can suffer substantial losses from fire through habitat destruction, changes in species composition, and less diversity [97]. Fire can sometimes help some open-land species by generating new openings, but it primarily disrupts native butterflies that are forest-dependent. In South Korea, for example, the delay of post-fire butterfly assemblages included the abundance of forest species and generalist butterflies increasing to greater levels and decreased abundance for forest specialists [97].

The immediate destruction of vegetation and forest structure dramatically increases further habitat loss [98]. In addition, in many regions, the loss of habitat may be compounded by climate warming, especially in tropical and subtropical regions, where there is potential for greater habitat degradation from climate warming [99]. The research found that in Vietnam, warming and habitat loss were shown to increase alterations in butterfly communities. Regimes of fires, including frequency and severity, are an important determinant of butterfly diversity and composition. Frequent fire can support savanna-adapted butterflies, whereas too infrequent fires can lead to forest invasion, which reduces the abundance of open grassy habitats and butterflies associated with these habitats [100].

The decline of the Oregon silverspot [*Speyeria zerene hippolyta*], which depends on coastal prairie habitats that are vulnerable to fire regimes, is a good example of this delicate balance. Therefore, understanding and managing fire regimes is an important aspect of the conservation of butterflies.

6. Living with the butterflies: Today and tomorrow

The term living with butterflies implies the present experience of coexistence with these insects and the prospect of having such positive experiences in the future, which might, however, not be predictable. Butterflies have frequently played a role, indicating a transformation and hope [101]. They also possess a typical evolution pathway of metamorphosis [egg, caterpillar, pupa, grown-up] [102]. It relates to the need of interacting with the biodiversity in terms of observing, documenting, and analyzing as a construct, which is the butterfly dimension. Descriptions, records, and counts of butterflies are all a part of this engagement so as to be able to inform evidence-based learning as well as adaptive management [103]. In the wake of today, it is important to live with butterflies by appreciating their beauty as well as their interaction with the environment and taking proactive steps toward their survival in the future. Defining their problems and establishing working conservation mechanisms, stakeholders can help these famous insects become resilient in a changing environment [101].

7. How to call butterflies back [interventions]

To effectively conserve butterflies, we need a complete set of actions ranging from habitat restoration and ongoing management to threat reduction [e.g., addressing invasive species and minimizing pesticides] and reducing larger-phased environmental stressors [104]. We need to take into account the ecology of butterflies broadly [host plant dynamics, microclimatic preferences, dispersal behavior, etc.] if we want to successfully conserve the butterflies [105].

One of the conservation actions we need to focus on is restoring degraded habitats, partly due to invasive alien plants, as invasive plants can directly [through their habitats] and indirectly [as invasive plants can change the quality of host plants, which can affect morphology and fecundity of adults] [106]. Manipulating habitat, such as planting back native plants or controlling *in situ* fire regimes, can all impact vital rates and survival. Another way to positively affect butterfly populations is grassland restoration. Although the lack of reasonable density estimates may limit the potential of some of these projects, there is still utility in using rigorous multiscale monitoring approaches in agroecosystems [107].

Butterflies have also been affected by human activities such as agriculture, the construction of cities, and pollution [8]. The damage can be minimized using more environmentally friendly farming and land-use practices, such as the reduction of chemical usage. The population of butterflies can also be maintained in cities by forming green areas such as parks and gardens [108]. Pollution should be reduced not only as a measure that would preserve butterflies but also other wildlife. Another excellent solution to conservation and creating awareness among the masses would be to get the people involved with citizen science [109]. Climate change creates further obstacles, altering habitats as well as the life cycles of butterflies. The development of wildlife corridors will enable all kinds of butterflies and other species to relocate as their habitats change [18].

An ideal combination of conservation tools should be used, which consists of the protection of the butterflies in their natural habitat and some targeted programs to preserve the butterfly population when the situation demands so.

By involving butterflies as umbrella species, it is possible to conserve other wildlife as well, and the traditional knowledge can provide the guidance that will be useful in the process of conserving other wildlife.

DOI: http://dx.doi.org/10.5772/intechopen.1011999

Author details

Unsar Naeem-Ullah*, Ayesha Tariq, Mirza Abdul Qayyum, Muhammad Asif Farooq, Farrukh Baig and Hafsah Ghaffar
Institute of Plant Protection, Muhammad Nawaz Shareef University of Agriculture, Multan, Pakistan

*Address all correspondence to: unsar.naeem@mnsuam.edu.pk

References

[1] Ramadhana F, Masy'ud B, Rahman DA. Diversity and colonization of butterfly [Lepidoptera] on Pasoso Island, Central Sulawesi. Media Konservasi. 2024;**28**(3):305-315

[2] Limbu R, Achint R. Diversity of butterfly in and around Vijaynagar of district Changlang, Arunachal Pradesh, India. International Journal of Tropical Insect Science. 2024;**44**(3):1319-1347

[3] Mackintosh A, Laetsch DR, Hayward A, Charlesworth B, Waterfall M, Vila R, et al. The determinants of genetic diversity in butterflies. Nature Communications. 2019;**10**(1):3466

[4] Basavarajappa S, Gopi KV, S. S. Butterfly species composition and diversity in a protected area of Karnataka, India. International Journal of Biodiversity and Conservation. 2018;**10**(10):432-443

[5] Dongmo MAK, Hanna R, Bonebrake TC. Enhancing scientific and community capacity to conserve central African Lepidoptera. Biological Conservation. 2023;**279**:109938

[6] Pinkert S, Farwig N, Kawahara AY, Jetz W. Global hotspots of butterfly diversity are threatened in a warming world. Nature Ecology and Evolution. 2025;**9**(5):789-800

[7] Yu XT, Yang FL, Da W, Li YC, Xi HM, Cotton AM, et al. Species richness of papilionidae butterflies [Lepidoptera: Papilionoidea] in the Hengduan mountains and its future shifts under climate change. Insects. 2023;**14**(3):259

[8] Wang WL, Suman DO, Zhang HH, Xu ZB, Ma FZ, Hu SJ. Butterfly conservation in China: From science to action. Insects. 2020;**11**(10):661

[9] Sonne J, Rahbek C. Idiosyncratic patterns of local species richness and turnover define global biodiversity hotspots. Proceedings of the National Academy of Sciences. 2024;**121**(3):e2313106121

[10] Bharati RS, Kumar S. Diversity, species richness and evenness of butterfly In Tarai region of Kumaon zone. The International Journal of Educational Research. 2024;**10**(1):84-87. Available from: https://ierj.in/journal/index.php/ierj/article/view/3263

[11] Gonzalez D, Pinto L, Sousa D, Oliveira I, Oliveira PS. Butterfly species richness and diversity on tourism trails of Northeast Portugal. Journal of Entomological Science. 2017;**52**(3):248-260

[12] Maes D, Gilbert M, Titeux N, Goffart P, Dennis RLH. Prediction of butterfly diversity hotspots in Belgium: A comparison of statistically focused and land use-focused models. Journal of Biogeography. 2003;**30**(12):1907-1920

[13] Kitahara M, Watanabe M. Diversity and rarity hotspots and conservation of butterfly communities in and around the Aokigahara woodland of Mount Fuji, Central Japan. Ecological Research. 2003;**18**(5):503-522

[14] Keerthika A, Parthiban KT. Multifunctional agroforestry landscapes: Augmenting butterfly biodiversity at foot hills of Nilgiris, India. International Journal of Tropical Insect Science. 2022;**42**(1):545-556

[15] Dollar JG, Riffell SK, Burger LW. Effects of managing semi-natural

grassland buffers on butterflies. Journal of Insect Conservation. 2013;**17**(3):577-590

[16] Pe'er G, Settele J. The rare butterfly *Tomares Nesimachus* [Lycaenidae] as a bioindicator for pollination services and ecosystem functioning in northern Israel. Israel Journal of Ecology and Evolution. 2008;**54**(1):111-136

[17] Cabral JS, Wiegand K, Kreft H. Interactions between ecological, evolutionary and environmental processes unveil complex dynamics of insular plant diversity. Journal of Biogeography. 2019;**46**(7):1582-1597

[18] Kemmerling LR, McCarthy AC, Brown CS, Haddad NM. Butterfly biodiversity increases with prairie strips and conservation management in row crop agriculture. Insect Conservation and Diversity. 2023;**16**(6):828-837

[19] Sharma K, Acharya BK, Sharma G, Valente D, Pasimeni MR, Petrosillo I, et al. Land use effect on butterfly alpha and beta diversity in the eastern Himalaya, India. Ecological Indicators. 2020;**110**:105605

[20] Rowe HI, Johnson B, Broatch J, Cruz TMP, Prudic KL. Winter rains support butterfly diversity, but summer monsoon rainfall drives post-monsoon butterfly abundance in the arid southwest of the US. Insects. 2023;**15**(1):5

[21] Maes D, Pardon W, Palmans G, Van Dyck H. The last of the maculineans: Can we save the emblematic Alcon blue butterfly Phengaris alcon under climate change when its habitat continues to deteriorate? Journal of Insect Conservation. 2024;**28**(5):1037-1049

[22] Habel JC, Schmitt T, Gros P, Ulrich W. Breakpoints in butterfly decline in Central Europe over the last century. Science of the Total Environment. 2022;**851**:158315

[23] Habel JC, Ulrich W, Biburger N, Seibold S, Schmitt T. Agricultural intensification drives butterfly decline. Insect Conservation and Diversity. 2019;**12**(4):289-295

[24] Inamine H, Ellner SP, Springer JP, Agrawal AA. Linking the continental migratory cycle of the monarch butterfly to understand its population decline. Oikos. 2016;**125**(8):1081-1091

[25] Crossley MS, Smith OM, Berry LL, Phillips-Cosio R, Glassberg J, Holman KM, et al. Recent climate change is creating hotspots of butterfly increase and decline across North America. Global Change Biology. 2021;**27**(12):2702-2714

[26] Dar AS, Javed Ansari M, Al Naggar Y, Hassan S, Nighat S, Burjes Zehra S, et al. Causes and reasons of insect decline and the way forward. In: Abdel Farag El-Shafie H, editor. Global Decline of Insects. London, UK: IntechOpen; 2022. Available from: https://www.intechopen.com/chapters/77656

[27] Wagner DL, Grames EM, Forister ML, Berenbaum MR, Stopak D. Insect decline in the Anthropocene: Death by a thousand cuts. Proceedings of the National Academy of Sciences. 2021;**118**(2):e2023989118

[28] United nations department for economic and social affairs. World Population Prospects 2024: Summary of Results. united nations: S.l.; 2025

[29] Sordello R, Flamerie De Lachapelle F, Livoreil B, Vanpeene S. Evidence of the environmental impact of noise pollution on biodiversity: A systematic map protocol. Environmental Evidence. 2019;**8**(1):8

[30] Adams JM, Kang C, June-Wells M. Are tropical butterflies more colorful? Ecological Research. 2014;**29**(4):685-691

[31] Vieira RRS, Pressey RL, Loyola R. The residual nature of protected areas in Brazil. Biological Conservation. 2019;**233**:152-161

[32] Spaniol RL, Mendonça MDS, Hartz SM, Iserhard CA, Stevens M. Discolouring the Amazon rainforest: How deforestation is affecting butterfly coloration. Biodiversity and Conservation. 2020;**29**(9-10):2821-2838

[33] Xing S, Au TF, Dufour PC, Cheng W, Landry Yuan F, Jia F, et al. Conservation of data deficient species under multiple threats: Lessons from an iconic tropical butterfly [Teinopalpus aureus]. Biological Conservation. 2019;**234**:154-164

[34] Adla K, Dejan K, Neira D, Dragana Š. Degradation of ecosystems and loss of ecosystem services. In: Prata JC, Ribeiro AI, Rocha-Santos TB, editors. One Health: Integrated Approach to 21st Century Challenges to Health. Academic Press; 2022. pp. 281-327

[35] Grames EM, Montgomery GA, Youngflesh C, Tingley MW, Elphick CS. The effect of insect food availability on songbird reproductive success and chick body condition: Evidence from a systematic review and meta-analysis. Ecology Letters. 2023;**26**(4):658-673

[36] Kucherov NB, Minor ES, Johnson PP, Taron D, Matteson KC. Butterfly declines in protected areas of Illinois: Assessing the influence of two decades of climate and landscape change. PLoS One. 2021;**16**(10):e0257889

[37] Forister ML, Halsch CA, Nice CC, Fordyce JA, Dilts TE, Oliver JC, et al. Fewer butterflies seen by community scientists across the warming and drying landscapes of the American west. Science. 2021;**371**(6533):1042-1045

[38] Davis AK, Schroeder H, Yeager I, Pearce J. Effects of simulated highway noise on heart rates of larval monarch butterflies, *Danaus plexippus*: Implications for roadside habitat suitability. Biology Letters. 2018;**14**(5):20180018

[39] Hågvar S, Ødegaard F, editors. Insect Conservation - Challenges and Possibilities in a Changing World. London, UK: IntechOpen; 2024. Available from: https://www.intechopen.com/books/1003265

[40] Maes D, Van Calster H, Herremans M, Van Dyck H. Challenges and bottlenecks for butterfly conservation in a highly anthropogenic region: Europe's worst case scenario revisited. Biological Conservation. 2022;**274**:109732

[41] Pallottini M, Goretti E, Argenti C, La Porta G, Tositti L, Dinelli E, et al. Butterflies as bioindicators of metal contamination. Environmental Science and Pollution Research. 2023;**30**(42):95606-95620

[42] Pignataro T, Lourenço GM, Cornelissen T. Shot down in grays: An integrative review and meta-analysis of the negative effects of urbanization on butterfly communities. Urban Ecosystem. 2025;**28**(3):116

[43] Deitsch JF, Huerta AC, Seymoure B. Butterflies at porch lights: Exploring nocturnal light visitation in butterflies using community science data from iNaturalist. Insect Conservative and Diversity. 2025;**18**(4):604-616

[44] Casner KL, Forister ML, O'Brien JM, Thorne J, Waetjen D, Shapiro AM.

Contribution of urban expansion and a changing climate to decline of a butterfly Fauna. Conservation Biology. 2014;**28**(3):773-782

[45] Guariento E, Rüdisser J, Fiedler K, Paniccia C, Stifter S, Tappeiner U, et al. From diverse to simple: Butterfly communities erode from extensive grasslands to intensively used farmland and urban areas. Biodiversity and Conservation. 2023;**32**(3):867-882

[46] Cardoso P, Barton PS, Birkhofer K, Chichorro F, Deacon C, Fartmann T, et al. Scientists' warning to humanity on insect extinctions. Biological Conservation. 2020;**242**:108426

[47] Warren MS, Maes D, Van Swaay CAM, Goffart P, Van Dyck H, Bourn NAD, et al. The decline of butterflies in Europe: Problems, significance, and possible solutions. Proceedings of the National Academy of Sciences. 2021;**118**(2):e2002551117

[48] Piccini I, Macrì M, Gea M, Dessì L, Bonetta S, Schilirò T, et al. Genotoxic effects of particulate matter on larvae of a common and widespread butterfly along an urbanization gradient. Ecotoxicology and Environmental Safety. 2023;**252**:114638

[49] Francis CD, Kleist NJ, Ortega CP, Cruz A. Noise pollution alters ecological services: Enhanced pollination and disrupted seed dispersal. Proceedings of the Royal Society B: Biological Sciences. 2012;**279**(1739):2727-2735

[50] Tassin De Montaigu C, Goulson D. Factors influencing butterfly and bumblebee richness and abundance in gardens. Science of Total Environment. 2024;**908**:167995

[51] Duque L, Steffan-Dewenter I. Air pollution: A threat to insect pollination. Frontiers in Ecology and the Environment. 2024;**22**(3):e2701

[52] Corke D. Are honeydew/sap-feeding butterflies [Lepidoptera: Rhopalocera] affected by particulate air-pollution? Journal of Insect Conservation. 1999;**3**(1):5-14

[53] Philips KH, Kobiela ME, Snell-Rood EC. Developmental lead exposure has mixed effects on butterfly cognitive processes. Animal Cognition. 2017;**20**(1):87-96

[54] Mupepele AC, Bruelheide H, Brühl C, Dauber J, Fenske M, Freibauer A, et al. Biodiversity in European agricultural landscapes: Transformative societal changes needed. Trends in Ecology and Evolution. 2021;**36**(12):1067-1070

[55] Sánchez-Bayo F, Wyckhuys KAG. Worldwide decline of the entomofauna: A review of its drivers. Biological Conservation. 2019;**232**:8-27

[56] Guderjan L, Habel JC, Schröder B, Schmitt T. Land-use pattern and landscape structure impact butterfly diversity and abundance in organic agroecosystems. Landscape Ecology. 2023;**38**(11):2749-2762

[57] Travis JMJ. Climate change and habitat destruction: A deadly anthropogenic cocktail. Proceedings of the Royal Society of London - Series B: Biological Sciences. 2003;**270**(1514):467-473

[58] Edwards CB, Schultz CB, Campbell SP, Fallon C, Henry EH, King KC, et al. Phenological constancy and management interventions predict population trends in at-risk butterflies in the United States. Journal of Applied Ecology. 2024;**61**(10):2455-2469

[59] Fox R, Dennis EB, Brown AF, Curson J. A revised red list of British butterflies. Insect Conservative and Diversity. 2022;**15**(5):485-495

[60] Dennis RLH, Dapporto L, Dover JW, Shreeve TG. Corridors and barriers in biodiversity conservation: A novel resource-based habitat perspective for butterflies. Biodiversity and Conservation. 2013;**22**(12):2709-2734

[61] Christian Habel J, Ulrich W, Gros P, Teucher M. Butterfly loss in urban landscapes and nature reserves. Journal for Nature Conservation. 2024;**78**:126562

[62] Kajzer-Bonk J, Nowicki P. Vanishing meadows — Quantitative analysis of factors driving population declines of endangered butterflies. Biological Conservation. 2023;**282**:110050

[63] Belsky J, Joshi NK. Assessing role of major drivers in recent decline of monarch butterfly population in North America. Frontiers in Environmental Science. 2018;**6**:86

[64] Kobiela ME, Snell-Rood EC. Genetic variation influences tolerance to a neonicotinoid insecticide in 3 butterfly species. Environmental Toxicology and Chemistry. 2020;**39**(11):2228-2236

[65] Yang LH, Postema EG, Hayes TE, Lippey MK, MacArthur-Waltz DJ. The complexity of global change and its effects on insects. Current Opinion in Insect Science. 2021;**47**:90-102

[66] Krishnan N, Zhang Y, Bidne KG, Hellmich RL, Coats JR, Bradbury SP. Assessing field-scale risks of foliar insecticide applications to monarch butterfly [*Danaus plexippus*] larvae. Environmental Toxicology and Chemistry. 2020;**39**(4):923-941

[67] Wan Zaki WM, Yahya MS, Norhisham AR, Sanusi R, Van Der Meer PJ, Azhar B. Agroforestry orchards support greater butterfly diversity than monoculture plantations in the tropics. Oecologia. 2023;**201**(3):863-875

[68] Asmah S, Ghazali A, Syafiq M, Yahya MS, Peng TL, Norhisham AR, et al. Effects of polyculture and monoculture farming in oil palm smallholdings on tropical fruit-feeding butterfly diversity. Agricultural and Forest Entomology. 2017;**19**(1):70-80

[69] Kukkonen JM, Von Numers M, Brommer JE. Conserving apollo butterflies: Habitat characteristics and conservation implications in Southwest Finland. Journal of Insect Conservation. 2024;**28**(6):1199-1210

[70] Kazemi H, Klug H, Kamkar B. New services and roles of biodiversity in modern agroecosystems: A review. Ecological Indicators. 2018;**93**:1126-1135

[71] Vujanović D, Arok M, Veselić S, Skendžić T, Andrić A, Đorđević A, et al. Butterfly community dynamics in a monoculture-dominated agricultural landscape. Ecological Entomology. 2025;**50**(2):360-372

[72] Wallisdevries MF, Van Swaay CAM, Plate CL. Changes in nectar supply: A possible cause of widespread butterfly decline. Current Zoology. 2012;**58**(3):384-391

[73] Balmori A. Electromagnetic radiation as an emerging driver factor for the decline of insects. Science of the Total Environment. 2021;**767**:144913

[74] Cucurachi S, Tamis WLM, Vijver MG, Peijnenburg WJGM, Bolte JFB, De Snoo GR. A review of the ecological effects of radiofrequency electromagnetic fields [RF-EMF]. Environment International. 2013;**51**:116-140

[75] Nyirenda VR, Namukonde N, Lungu EB, Mulwanda S, Kalezu K, Simwanda M, et al. Effects of phone mast-generated electromagnetic radiation gradient on the distribution of terrestrial birds and insects in a savanna protected area. Biologia [Bratisl]. 2022;**77**(8):2237-2249

[76] Starlin NM, Rani SA. Butterfly diversity and host plant dynamics in a semi-urban ecosystem: Insights from the Nesamony memorial Christian College campus, Marthandam, Tamil Nadu, India. Uttar Pradesh Journal of Zoology. 2024;**45**(23):158-168

[77] Nieto-Sánchez S, Gutiérrez D, Wilson RJ. Long-term change and spatial variation in butterfly communities over an elevational gradient: Driven by climate, buffered by habitat. Diversity and Distributions. 2015;**21**(8):950-961

[78] Oliver TH, Thomas CD, Hill JK, Brereton T, Roy DB. Habitat associations of thermophilous butterflies are reduced despite climatic warming. Global Change Biology. 2012;**18**(9):2720-2729

[79] Habel JC, Teucher M, Gros P, Schmitt T, Ulrich W. Land use and climate change affects butterfly diversity across northern Austria. Landscape Ecology. 2021;**36**(6):1741-1754

[80] Wepprich T, Henry E, Haddad NM. Voltinism shifts in response to climate warming generally benefit populations of Multivoltine butterflies. Ecology Letters. 2025;**28**(4):e70018

[81] Wallisdevries MF, Van Swaay CAM. Global warming and excess nitrogen may induce butterfly decline by microclimatic cooling. Global Change Biology. 2006;**12**(9):1620-1626

[82] Chen XM, Sharma A, Liu H. The impact of climate change on environmental sustainability and human mortality. Environments. 2023;**10**(10):165

[83] Oliveira IF, Baccaro FB, Werneck FP, Haugaasen T. Seasonal flooding decreases fruit-feeding butterfly species dominance and increases spatial turnover in floodplain forests of Central Amazonia. Ecology and Evolution. 2023;**13**(1):e9718

[84] Fies R, Rabl D, Schulze CH, Fiedler K. Summer floods shape meadow butterfly communities in a floodplain nature reserve in Central Europe. Journal of Insect Conservation. 2016;**20**(3):433-445

[85] Graham L, Gaulton R, Gerard F, Staley JT. The influence of hedgerow structural condition on wildlife habitat provision in farmed landscapes. Biological Conservation. 2018;**220**:122-131

[86] Steele C, Ragonese IG, Majewska AA. Extent and impacts of winter breeding in the north American monarch butterfly. Current Opinion in Insect Science. 2023;**59**:101077

[87] Hayamizu M, Nakahama N, Ohwaki A, Kinoshita G, Uchida Y, Koyama N, et al. Effect of mowing on population maintenance of the endangered silver-studded blue butterfly, Plebejus subsolanus [Lepidoptera: Lycaenidae], throughout its life cycle in Japan. Journal of Insect Conservation. 2024;**28**(3):437-448

[88] Crous CJ, Pryke JS, Samways MJ. Conserving a geographically isolated *Charaxes* butterfly in response to habitat fragmentation and invasive alien plants. Koedoe. 2015;**57**(1):9

[89] Aota T, Ashizawa K, Mori H, Toda M, Chiba S. Detection of Anolis carolinensis

using drone images and a deep neural network: An effective tool for controlling invasive species. Biological Invasions. 2021;**23**(5):1321-1327

[90] Lourenço GM, Luna P, Guevara R, Dáttilo W, Freitas AVL, Ribeiro SP. Temporal shifts in butterfly diversity: Responses to natural and anthropic forest transitions. Journal of Insect Conservation. 2020;**24**(2):353-363

[91] Bonifacino M, Pasquali L, Sistri G, Menchetti M, Santini L, Corbella C, et al. Climate change may cause the extinction of the butterfly Lasiommata petropolitana in the Apennines. Review. 2022;**26**:959-972. Available from: https://www.researchsquare.com/article/rs-1573572/v1

[92] Löffler F, Grüneberg C, Drung M, Freienstein FM, Helbing F, Holtmann L, et al. Different environmental conditions in lowlands and uplands highlight challenges for butterfly conservation in Central Europe. Biological Conservation. 2023;**281**:110034

[93] Oki K, Soga M, Amano T, Koike S. Abandonment of ski pistes impoverishes butterfly communities. Biodiversity and Conservation. 2022;**31**(8-9):2083-2096

[94] Kyriacou CP, Rosato E. Genetic analysis of cryptochrome in insect magnetosensitivity. Frontiers in Physiology. 2022;**13**:928416

[95] Guerra PA, Matter S. Changes in the geomagnetic field has little effect on the overwintering range of eastern north American fall migratory monarch butterflies [Danaus plexippus], indicating a lack of an innate magnetic map sense for navigation. [preprint]. Research Square. Review. 3 Feb 2020. Available from: https://www.researchsquare.com/article/rs-13021/v1

[96] Gorobets O, Gorobets S, Polyakova T, Zablotskii V. Modulation of calcium signaling and metabolic pathways in endothelial cells with magnetic fields. Nanoscale Advances. 2024;**6**(4):1163-1182

[97] Kim SS, Kwon TS. Changes in butterfly assemblages and increase of open-land inhabiting species after forest fires. Journal of Asia-Pacific Biodiversity. 2018;**11**(1):39-48

[98] Palm EC, Suitor MJ, Joly K, Herriges JD, Kelly AP, Hervieux D, et al. Increasing fire frequency and severity will increase habitat loss for a boreal forest indicator species. Ecological Applications. 2022;**32**(3):e2549

[99] Bonebrake TC, Pickett EJ, Tsang TPN, Tak CY, Vu MQ, Vu LV. Warming threat compounds habitat degradation impacts on a tropical butterfly community in Vietnam. Global Ecology and Conservation. 2016;**8**:203-211

[100] Stone ZL, Maron M, Tasker E. Reduced fire frequency over three decades hastens loss of the grassy forest habitat of an endangered songbird. Biological Conservation. 2022;**270**:109570

[101] Deaton CCM, Nicholson H. Interacting with butterflies. Teaching Children Mathematics. 2015;**22**(5):280-281

[102] Yasmin R, Das A, Rozario LJ, Islam ME. Butterfly detection and classification techniques: A review. Intelligent Systems with Applications. 2023;**18**:200214

[103] Pandian A. Butterfly crossings. Environmental Humanities. 2022;**14**(2):438-456

[104] Mcintire EJB, Schultz CB, Crone EE. Designing a network for butterfly habitat restoration: Where individuals, populations and landscapes interact. Journal of Applied Ecology. 2007;**44**(4):725-736

[105] Sivakoff FS, Morris WF, Aschehoug ET, Hudgens BR, Haddad NM. Habitat restoration alters adult butterfly morphology and potential fecundity through effects on host plant quality. Ecosphere. 2016;**7**(11):e01522

[106] Schultz CB, Ferguson JA. Demographic costs and benefits of herbicide-based restoration to enhance habitat for an endangered butterfly and a threatened plant. Restoration Ecology. 2020;**28**(3):564-572

[107] Kral KC, Hovick TJ, Limb RF, Harmon JP. Multi-scale considerations for grassland butterfly conservation in agroecosystems. Biological Conservation. 2018;**226**:196-204

[108] Miao B, Peng Y, Yang D, Kubota Y, Economo EP, Liu C. Climate and land-use interactively shape butterfly diversity in tropical rainforest and savanna ecosystems of southwestern China. Insect Science. 2021;**28**(4):1109-1120

[109] Bubová T, Vrabec V, Kulma M, Nowicki P. Land management impacts on European butterflies of conservation concern: A review. Journal of Insect Conservation. 2015;**19**(5):805-821

Chapter 2

Integrating Morphological, Molecular and Whole Genome (Mitogenomic) Approaches in Indian Lepidoptera

Aquib Majeed, Tabraq Ali, Zahid Hussian and Sajad Hussain Parey

Abstract

India, with its extensive and diverse biogeographical regions, supports a remarkably rich Lepidopteran diversity, particularly in the Himalayas, Western Ghats and Northeastern areas. Despite this vast biodiversity, taxonomic studies on Indian Lepidoptera have primarily relied on traditional morphological approaches, with limited integration of molecular and genomic techniques. The lack of comprehensive genomic datasets for Indian Lepidoptera presents a major hurdle in resolving taxonomic ambiguities, particularly in cryptic and endemic species. Moreover, challenges such as habitat fragmentation, climate change and human induced disturbances have intensified conservation concerns, emphasizing the need for more precise taxonomic evaluations. Recent advancements have focused on addressing these limitations by incorporating morphological, molecular and whole genome (Mitogenome) methodologies into Lepidopteran research in India. The sequencing of complete mitochondrial genomes and nuclear markers for selected species has provided valuable insights into phylogeography and evolutionary relationships. However, large scale genomic initiatives are yet to gain substantial traction in India. This chapter shall provide a detailed review of Lepidoptera taxonomy, emphasizing integration of morphological, molecular and whole genome approaches. Special attention shall be given to Indian Lepidoptera, discussing existing knowledge gaps, recent progress and future directions. By combining classical taxonomy with molecular and genomic tools, this chapter aims to refine taxonomy and conservation efforts in India.

Keywords: conservation, database, lepidoptera, mitogenome, taxonomy

1. Introduction

Lepidoptera, one of the most diverse insect orders, comprises over 160,000 described species of butterflies and moths worldwide [1], with India being home to an estimated 15,000–20,000 species due to its rich biogeographical diversity [2]. These insects play vital ecological roles as pollinators, herbivores and prey, and serve as key bioindicators of environmental change [3]. However, accurate species identification and phylogenetic understanding of Indian Lepidoptera remain challenging due to the presence of cryptic species, sexual dimorphism and morphological plasticity. Traditional taxonomy based solely on morphological traits often falls short in resolving complex species boundaries [4].

Traditional classification of Lepidoptera based on morphological features such as wing patterns, genitalia structures and venation has played a foundational role in taxonomic studies. However, reliance on morphology alone often proves inadequate in the face of cryptic species, sexual dimorphism and convergent evolution [5].

In recent years, integrative taxonomy combining morphological analysis with molecular and genomic tools has emerged as a robust framework to overcome such limitations. DNA barcoding has shown promise in species identification and biodiversity assessment, especially when employing the mitochondrial cytochrome c oxidase I (COI) gene. Beyond barcoding, mitogenomic (whole mitochondrial genome) approaches are gaining prominence for their enhanced resolution in phylogenetics, population genetics and evolutionary studies [6]. Rich genetic data from mitogenomes, such as gene order, codon use and regulatory region variants, can be used to better understand historical biogeography and lineage divergence. Integrating morphological, molecular (e.g., COI, ITS and 28S) and mitogenomic data offers a comprehensive toolkit for accurate species delineation, discovery of novel taxa and clarification of evolutionary relationships within Indian Lepidoptera [7]. The purpose of this study is to emphasize the synergistic value of combining these methodologies to advance Lepidopteran research in India, thereby contributing to global efforts in biodiversity documentation and management.

Mitogenome has several unique features like maternal inheritance, small genome size, faster evolution rate, low or absence of homologous recombination, evolutionarily conserved gene products and richness in genetic polymorphism, which makes it a potential marker for barcoding, phylogeography and phylogenetic studies. By elucidating evolutionary models and substitution rates/patterns that differ over time and between sequences, it may be useful in molecular evolutionary research. Because of their numerous genome-level characteristics, such as gene position, content, secondary RNA structures and regulatory regions, complete mitogenomes provide more relevant evolutionary models than individual genes [8].

Lepidopteran taxonomy has changed with the advent of molecular tools, especially DNA barcoding using the mitochondrial cytochrome c oxidase I (COI) gene. Molecular data offer improved resolution in species identification, reveal hidden biodiversity and aid in phylogenetic reconstructions. More recently, complete mitochondrial genome (mitogenome) sequencing has emerged as a powerful tool, providing comprehensive genetic information beyond single-gene markers [9]. The integration of morphological, molecular and mitogenomic data commonly termed as integrative taxonomy has gained traction as a robust approach to resolve taxonomic ambiguities and understand evolutionary relationships more precisely. Morphological identification has long been the cornerstone of Lepidopteran taxonomy. Key diagnostic traits include wing coloration and patterns, antennae shape and size, genitalia structures, venation and scale arrangements. While morphology remains essential, it has limitations, including morphologically similar (cryptic) species and intraspecific variation due to developmental or environmental factors, degraded specimens lacking identifiable features [10].

The introduction of DNA barcoding revolutionized insect taxonomy. In Lepidoptera, the mitochondrial COI gene is widely used for species identification.

Beyond COI, nuclear markers such as Internal Transcribed Spacer (ITS), Elongation Factor-1 alpha (EF-1α) and 28S rRNA are used for deeper phylogenetic studies. DNA barcoding projects in India, including contributions to the Barcode of Life Data System (BOLD), have cataloged many Lepidopteran species [11]. However, barcoding alone has limitations; Pseudogenes may interfere with results, and there is limited resolution in closely related taxa.

2. Whole mitochondrial genome (mitogenomic) approaches

Mitogenomics offers a more comprehensive dataset for taxonomic and phylogenetic research. Mitochondrial genomes are compact (15–20 kb), maternally inherited and relatively conserved in gene content and order. Mitogenomic data have been used to resolve deep-level relationships within superfamilies like Noctuoidea and Gelechioidea [12]. In India, mitogenomic studies are still emerging but show great potential in refining classification and discovering new taxa. Integrative taxonomy, by combining morphological, molecular and mitogenomic data, offers a powerful approach to unravel the complex biodiversity of Indian Lepidoptera. It enhances species identification accuracy, supports conservation planning and enriches our understanding of evolutionary histories. As technology becomes more accessible and collaborative efforts strengthen, integrative methods will play an increasingly pivotal role in documenting and preserving India's rich Lepidopteran fauna. Mitogenomics, has rapidly evolved as a powerful tool in systematics, offering high-resolution insights into the taxonomy and evolutionary relationships of various organisms [13]. In Lepidoptera, the second largest order of insects, with over 160,000 described species globally, mitogenomic data have proven particularly valuable. Mitochondrial genomes are relatively small (typically 15–20 kb), circular in structure and contain 37 genes, including 13 protein-coding genes, 22 tRNAs and 2 rRNAs. Due to their maternal inheritance, low recombination rate and relatively conserved gene arrangement, mitochondrial genomes are ideal for phylogenetic studies at both shallow and deep evolutionary levels [14].

3. Global advances in mitogenomics

Worldwide, mitogenomic studies have significantly reshaped our understanding of Lepidopteran evolution. Extensive research on superfamilies such as Noctuoidea, Papilionoidea and Gelechioidea has led to refined classifications, detection of cryptic species and resolution of long-standing taxonomic ambiguities. For instance, the use of complete mitochondrial genomes has clarified relationships in groups where traditional markers (like COI) failed to resolve phylogenetic positions due to saturation or convergence [15]. Additionally, mitogenomics has been increasingly integrated with nuclear genomic data, morphological characters and ecological traits to establish a comprehensive and robust framework for species delimitation, an approach known as integrative taxonomy. This method not only enhances taxonomic accuracy but also supports broader goals such as conservation biology, biogeography and evolutionary ecology [16].

Recent advances in genomic technologies such as whole genome sequencing (WGS), RNA sequencing (RNA-seq) and CRISPR/Cas9-based gene editing have significantly enhanced our ability to unravel the genetic architecture underlying various biological traits, including insecticide resistance in Lepidoptera. The availability of high-quality whole genome assemblies for model species such as *Bombyx mori* has been particularly instrumental in elucidating the molecular basis of adaptive responses in Lepidopteran taxa [17]. These genomic resources provide a robust platform for identifying candidate genes, molecular pathways and regulatory networks associated with resistance mechanisms. Despite these technological strides, the practical implementation of gene-editing technologies in pest management faces substantial challenges. The potential for off-target genomic alterations, strict regulatory frameworks and ecological hazards connected to the discharge of genetically modified organisms into the environment are major concerns. Furthermore, the high genetic variability and phenotypic plasticity exhibited by Lepidopteran pest populations across India pose additional hurdles for developing durable resistance management strategies [18].

In this context, the integration of functional genomics with morphological and mitogenomic data provides a more holistic framework for understanding resistance evolution and pest adaptability. This chapter synthesizes current knowledge on the molecular mechanisms of resistance in Indian Lepidoptera, while also evaluating the potential of advanced genomic and gene-editing tools to inform integrative taxonomy, evolutionary studies and pest control initiatives. By aligning morphological diagnostics with high-resolution molecular and genomic data, this integrative approach not only enhances species delimitation but also strengthens the foundation for ecologically sound and sustainable pest management strategies.

4. The Indian landscape: Potentials and progress

In India, a mega-diverse country with over 15,000 known species of Lepidoptera [2], the application of mitogenomic tools is still in its infancy. Most molecular taxonomic studies so far have relied heavily on partial mitochondrial markers like COI for DNA barcoding. However, such single-locus approaches often fall short in resolving deeper evolutionary relationships or distinguishing between closely related species complexes. Recent pioneering studies have started sequencing and analyzing complete mitochondrial genomes from Indian Lepidoptera, offering new insights into endemic lineages, cryptic diversity and phylogeographic patterns across the country's varied biogeographical zones from the Himalayas and Western Ghats to the north-eastern biodiversity hotspot [19]. These efforts are crucial in a region facing rapid habitat loss and climate change.

Institutes like the Zoological Survey of India (ZSI), National Centre for Biological Sciences (NCBS) and several universities are now collaborating on mitogenomic research. With the increasing availability of high-throughput sequencing platforms and bioinformatics tools, India is poised to contribute significantly to global mitogenomic databases and comparative phylogenetics.

5. Integrative taxonomy: A way forward

The future of Lepidopteran taxonomy, both globally and in India, lies in integrative taxonomy, which combines classical morphological data with molecular (COI, nuclear genes) and mitogenomic evidence. This holistic approach enhances the reliability of species identifications, especially in groups where traditional taxonomy has been challenged by convergent morphology or intraspecific variation. In India, integrative taxonomy is especially relevant for: (a) Discovering and describing new species, especially in underexplored regions like the northeast Himalayas or the Western Ghats. (b) Revising outdated classifications, many of which are still based solely on morphology from colonial-era records. (c) Supporting conservation action, by identifying evolutionarily significant units and prioritizing them in protected area networks. (d) Informing pest management strategies, particularly for agriculturally important groups like Pyraloidea or Noctuoidea [20].

6. Case study

In January 2022, Kalawate et al. [21] published a pioneering study on *Asota ficus* (Fabricius, 1775) from India, combining meticulous morphological analysis with mitochondrial DNA barcoding to resolve its taxonomic status. The morphological assessment encompassed external traits such as wing coloration, fasciculated antennae in males and distinctive genitalia structures for both sexes based on voucher specimens collected from various Indian locales. In tandem, the researchers generated the first COI barcode of the species from Indian populations and deposited it alongside the voucher specimen, enabling direct comparison with global *A. ficus* sequences. The genetic analysis revealed only shallow divergence among barcodes, supporting the morphological identification and affirming that the collected Indian samples belong to a genetically cohesive species. This integrative methodology not only anchors a reliable reference point for *A. ficus* in barcode repositories but also highlights the power of combining classical taxonomy with molecular tools to clarify species boundaries, detect potential cryptic diversity and strengthen biodiversity assessments in Indian Lepidoptera. The other representative studies summarized in **Table 1** illustrate the increasing use of integrative approaches combining morphological and molecular data for resolving complex taxonomic and phylogenetic issues in Lepidoptera.

S. No	Summary of Study	Target Group (Taxa)	Approach Used	References
1.	Mitochondrial genome characterization of the Indian tasar *Antheraea mylitta* silkworm and its phylogenetic implications	Saturniidae (Bombycoidea)	Mitogenomics + Morphology	Kim et al. [22]
2.	Molecular and morphological divergence in Western Himalayan butterfly species(4 species)	Papilionoidea (Nymphalidae, Pieridae)	COI Barcoding + Morphological Keys	Singh et al. [23]
3.	Morphological characterization and mtDNA barcode of *Asota ficus* tiger moth from India	Lepidoptera: Noctuoidea: Erebidae: Aganainae	Morphology + Genitalia + COI barcoding	Kalawate et al. [21]
4.	Complete mitogenome sequencing of Muga silkworm along with its comparative analysis across Lepidoptera	Bombycidae (Bombycoidea)	Whole mitogenome + comparative phylogenetics	Singh et al. [8]
5.	Use of ITS2 structure alongside COI and morphology to refine the blue butterfly phylogeny	*Agrodiaetus* (Lepidoptera: Lycaenidae: *Polyommatus*)	Morphology + COI + ITS2 sequencing	Wiemers et al. [24]

Table 1.
Representative studies using integrative morphological and molecular approaches in Lepidoptera taxonomy and phylogeny.

7. Looking ahead

As sequencing technologies become more cost-effective and widespread and as collaborative frameworks (such as global DNA barcoding initiatives and regional biodiversity missions) grow stronger, mitogenomics will become increasingly integral to the study of Lepidoptera. For India, investing in training, infrastructure and policy support for integrative taxonomic research will be crucial. Therefore, mitogenomics and integrative taxonomy are transforming how we understand, document and conserve biodiversity [20]. For Lepidoptera, a group both ecologically vital and esthetically celebrated, these tools offer unprecedented opportunities to explore evolutionary histories and support long-term ecological resilience, both in India and across the globe.

8. Methodology and experimental procedures

8.1 Isolation of genomic DNA, library construction and sequencing

The DNeasy Blood and Tissue Kit (Qiagen, Germany) can be used to extract high-molecular-weight genomic DNA from individual male specimens following the manufacturer's standard operating procedure. Agarose gel electrophoresis and a NanoDrop spectrophotometer (Thermo Scientific, USA) are frequently employed in the evaluation of DNA integrity and purity. Only samples meeting stringent quality thresholds should be selected for sequencing. To ensure comprehensive genome coverage, whole-genome sequencing can be carried out utilizing both short-read and long-read technologies. The NEXTflex Rapid DNA Sequencing Kit (PerkinElmer, USA) enables the preparation of short-read libraries with insert sizes of 250–550 and 300–700 bp. The Agilent TapeStation system (Agilent Technologies, USA) is used to evaluate the fragment size distribution of these libraries. The Nextera Mate Pair Library Preparation Kit (Illumina Inc., USA) can be used to create paired-end Illumina mate-pair libraries with larger insert sizes (5–7 and 7–10 kb) in order to facilitate robust genome assembly. Additionally, a PacBio SMRTbell library can be prepared for long-read sequencing, enhancing scaffold continuity and resolving complex genomic regions.

All sequencing procedures are typically conducted at specialized facilities, such as Genotypic Technology Pvt. Ltd., Bangalore, India, employing high-throughput Illumina and PacBio platforms. The integration of short- and long-read datasets ensures high-quality de novo genome assembly, suitable for downstream phylogenomic, comparative and functional analyses within the broader framework of Indian Lepidoptera research.

8.2 Phylogenetic analyses

To investigate the evolutionary relationships among selected Lepidopteran taxa, comprehensive phylogenetic analyses are conducted using mitogenomic datasets generated through a combination of MEGA X and PhyloSuite v1.2.1. Five distinct datasets are constructed to capture different levels of phylogenetic signal:

PCG12: First and second codon positions of all 13 protein-coding genes (PCGs);

PCG123: All codon positions of the 13 PCGs;

PCG12R: First and second codon positions of the 13 PCGs plus 24 mitochondrial RNA genes (22 tRNAs +2 rRNAs);

PCG123R: All codon positions of 13 PCGs along with 24 RNA genes;

PCGAA: Translated amino acid sequences derived from the 13 PCGs [25].

8.3 Maximum likelihood (ML) analyses

Felsenstein first introduced maximum likelihood (ML) in the early 1980s. **Figure 1** illustrates the primary steps involved in building an evolutionary tree with this approach.

First, an appropriate evolutionary model is selected depending on the characteristics of the sequencing data. The JC69, K80, TN93, HKY85 and GTR are examples of common models. While TN93 calculates base frequencies from the data and takes into consideration varying transition and transversion rates [27] the JC69 model assumes equal substitution rates among all nucleotides [26]. Different base frequencies and substitution rates are supported under the GTR model.

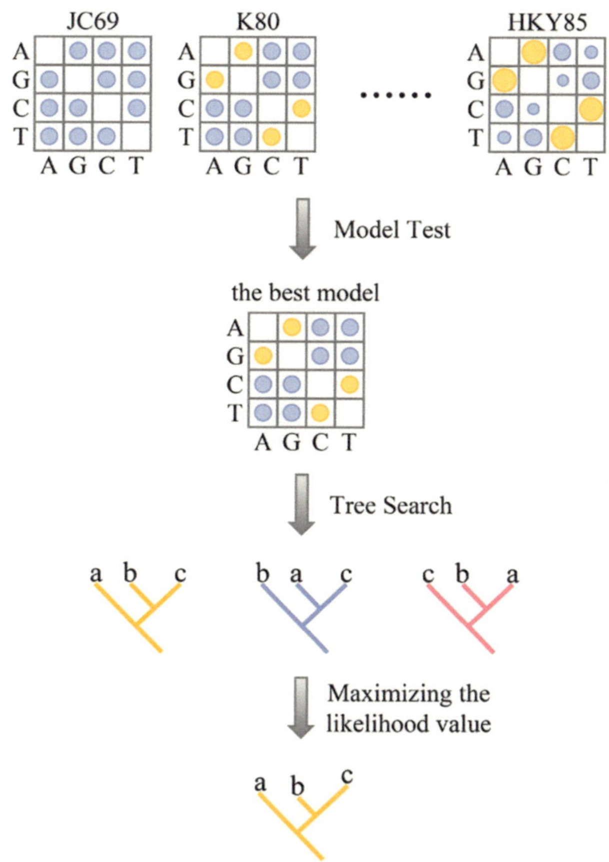

Figure 1.
Models like JC69 (Jukes-Cantor 1969), K80 (Kimura 1980) and HKY85 (Hasegawa-Kishino-Yano 1985) are essential to the maximum likelihood approach for phylogenetic tree construction; species are designated as a, b and c [26].

Then, using conventional numerical techniques, a tree space search is carried out, optimizing branch lengths and substitution parameters for every topology in order to maximize the likelihood value. Topologies having the highest maximum likelihood (ML) value are considered ideal trees. To determine the optimal tree, the topology with the highest maximum likelihood (ML) value is chosen. Although it is preferable to evaluate every possible tree, the number of hypothetical trees rapidly increases as the number of taxa climbs. Accordingly, heuristic approaches are necessary for larger datasets, while exhaustive searches are only practical for small datasets [28].

Likelihood methods, grounded in explicit models, reduce systematic errors like long-branch attraction compared to parsimony methods. However, their complex models increase computational demands. Maximum likelihood remains statistically consistent, robust and effective for comparing trees and utilizing full sequence data within a statistical framework.

8.4 Bayesian inference (BI) analyses

Genomic sequence analysis was revolutionized in the 1990s when Bruce Rannala and Ziheng Yang developed Bayesian inference (BI) for phylogenetic analysis [29]. In contrast to maximum likelihood (ML) techniques, Bayesian methods quantify parameter uncertainty using statistical distributions. In contrast to maximum likelihood (ML) methods, Bayesian methods quantify parameter uncertainty using statistical distributions. Choosing a suitable evolutionary model and establishing previous knowledge (such as tree topology and branch lengths) based on expert knowledge are the first steps in the process, as illustrated in **Figure 2** [30].

Random samples are then taken from the posterior distribution using Markov Chain Monte Carlo (MCMC) sampling. Every sample creates a Markov chain that converges to a stationary distribution that matches the posterior of the phylogenetic tree. Metropolis-Hastings, Metropolis-coupled MCMC and Larget and Simon's LOCAL algorithm are examples of common MCMC algorithms. Their posterior probabilities are assessed based on the proportion of sampled trees with branches and trees; the optimal tree is determined by its highest posterior probability [26]. Bayesian inference offers advantages such as faster computation on large datasets and the ability to assess tree confidence through posterior probabilities.

8.5 Mitogenome annotation and data analyses

Mitogenome annotation and analysis in Lepidoptera have become essential tools for understanding evolutionary relationships, gene rearrangements and adaptation. Typically, Lepidopteran mitogenomes are circular molecules, as shown in **Figure 3** of ~15–16 kb, containing 37 genes, 13 protein-coding genes (PCGs), 22 tRNAs, 2 rRNAs and a non-coding A + T-rich control region. Annotation involves assembling mitogenomes from genomic or transcriptomic data, identifying gene boundaries using tools like MITOS and tRNAscan-SE, and confirming structural features such as the frequent loss of the DHU arm in trnS1. The control region is characterized by motifs like ATAGA and microsatellite repeats. Comparative analyses reveal a strong A + T nucleotide bias, skewed codon usage (with high frequency of UUA for leucine) and occasional rearrangements of tRNA genes that serve as phylogenetic markers. Continuous-time Markov processes model nucleotide substitutions, and phylogenetic analyses often employ concatenated PCG datasets using Maximum Likelihood and Bayesian methods to resolve deep evolutionary relationships [31]. (Balakrishnan

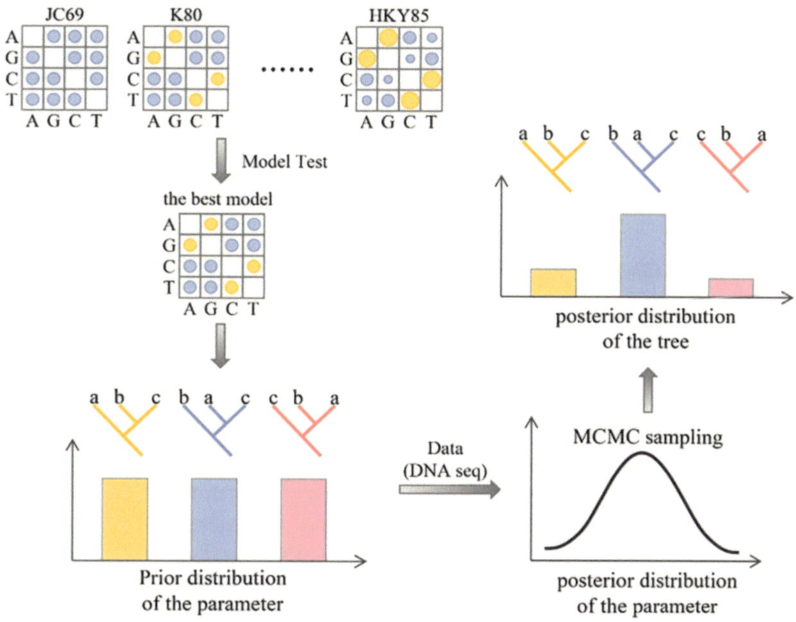

Figure 2.
Bayesian inference for phylogenetic tree construction utilizes models such as JC69 (Jukes-Cantor 1969), K80 (Kimura 1980) and HKY85 (Hasegawa-Kishino-Yano 1985), with MCMC (Markov chain Monte Carlo) for estimating tree probabilities; species are denoted as a, b and c [26]. For nucleotide substitution, the majority of models use continuous-time Markov processes (CTMPs), which possess the Markov property, which claims that future states are only dependent on the present state and not the past. The posterior probability distribution of parameters is obtained by applying the Bayes theorem. It integrates the likelihood of sequence data with parameter priors [28].

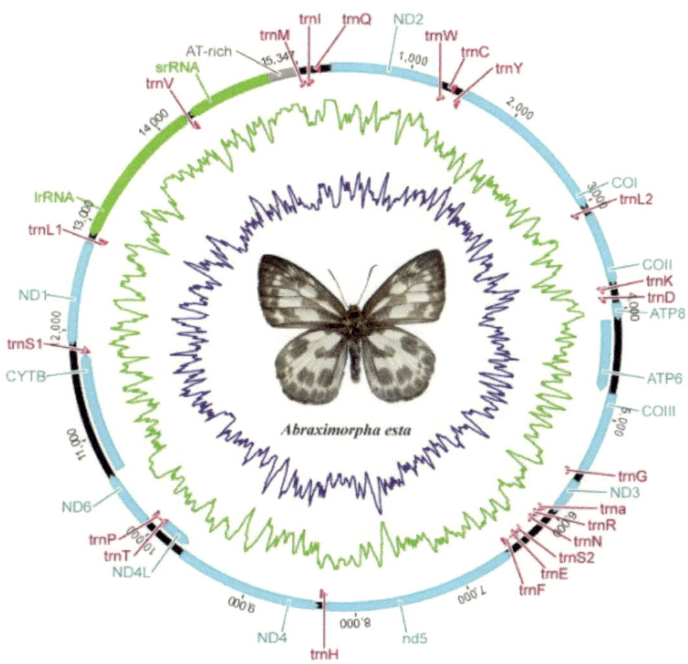

Figure 3.
A circular representation of the mitochondrial genome of Abraximorpha esta *[32].*

et al., 2024). Additionally, research has found evidence of adaptive evolution in mitochondrial genes, particularly in species that live in harsh environments. These mitogenomic insights not only improve taxonomic resolution across Lepidoptera but also shed light on mitochondrial genome evolution and function [6].

9. Conclusions

9.1 Reassessing the taxonomic landscape

Lepidoptera, the second largest insect order globally, represents an unparalleled diversity of form and function, particularly in India, a country with immense ecological heterogeneity. With an estimated 15,000–20,000 species, Indian Lepidoptera constitute a significant portion of the global moth and butterfly fauna [33]. Despite this richness, the taxonomic clarity and phylogenetic positioning of many Indian taxa remain unresolved due to overlapping morphologies, the presence of cryptic species, and outdated classifications based solely on colonial-era morpho-taxonomic schemes [34]. Conventional taxonomy, which has traditionally relied on morphological features such as wing coloration, venation and genitalia structures, while foundational, is limited in resolving fine-scale or deep evolutionary relationships. The emergence of DNA barcoding, primarily using the mitochondrial COI gene, partially addressed these issues by enabling faster and more reliable species identifications. However, single-gene markers often lack the resolution necessary for clarifying complex species boundaries, detecting recent divergence or resolving long-standing taxonomic ambiguities. This becomes particularly problematic in speciose groups such as Noctuoidea, Pyraloidea and Gelechioidea, where intraspecific variability and interspecific mimicry are rampant [5].

The integration of whole mitochondrial genome data (mitogenomics) represents a logical and transformative extension of the molecular taxonomy movement. This approach not only supplements traditional barcoding methods but also offers a phylogenetically rich dataset that improves resolution at both shallow and deep taxonomic levels. Through parameters like gene order rearrangements, codon usage bias, RNA secondary structures and control region variability, mitogenomes deliver multi-dimensional insights into Lepidopteran diversification.

9.2 Mitogenomics as a transformative tool

Mitogenomes are compact, maternally inherited and evolve rapidly compared to nuclear genomes, making them ideal for comparative studies. Typically, insect mitogenomes consist of 13 protein-coding genes, 22 tRNAs, 2 rRNAs and a control region. What makes the mitochondrial genome uniquely valuable is its combination of structural conservation and sequence-level variability features that together enable both conserved marker utility and lineage-specific differentiation. Globally, mitogenomic research has already revolutionized our understanding of Lepidopteran phylogeny [35]. Refined classifications within superfamilies such as Noctuoidea, Papilionoidea and Bombycoidea have emerged as a direct result of mitogenomic insights. These studies have corrected erroneous groupings based on superficial similarities and have shed light on deep evolutionary divergences not evident through morphology or COI alone [36].

In India, however, this tool remains significantly underutilized. Only a handful of complete mitogenomes have been sequenced from Indian Lepidoptera, and those that exist are largely restricted to economically significant taxa like *Bombyx mori* or major pest species. This data gap not only hampers global comparative studies but also limits national biodiversity assessments. With accelerating habitat degradation and climate-induced shifts in species distributions, there is an urgent need to fill these molecular knowledge voids [37].

9.3 The Indian opportunity: Bridging the molecular divide

India's geographic and ecological diversity from the alpine zones of the western Himalayas and the Eastern Himalayan biodiversity hotspot, to the Western Ghats wet forests and arid Thar desert offers an exceptional setting for studying evolutionary radiations and cryptic speciation. Yet, the current taxonomic output often fails to capture this diversity at a genomic resolution [38].

Encouragingly, Indian institutions such as the Zoological Survey of India (ZSI), National Centre for Biological Sciences (NCBS) and several state universities have begun laying the groundwork for such studies. With access to high-throughput sequencing platforms, genome annotation tools and trained personnel, India is now equipped to contribute substantially to global mitogenomic databases [39]. Mitogenomic data combined with nuclear markers (e.g., 28S rRNA, EF-1α and ITS) constitute a cornerstone of integrative taxonomy, which serves as the gold standard in contemporary systematic biology. This tripartite approach (morphology + nuclear genes + mitogenome) is not just an academic exercise; it holds practical implications for pest management, conservation prioritization and understanding disease-vector ecology.

9.4 Future directions: Building a national mitogenome initiative

To launch a "National Mitogenome Initiative for Lepidoptera", a systematic large-scale sequencing effort is proposed, targeting representative taxa from all major families and biogeographic zones across India. Such an undertaking:

- Generate a robust reference library of mitogenomes to support species identifications and biodiversity assessments.
- Enable comparative phylogeographic studies across ecological gradients (e.g., altitudinal clines in the Himalayas).
- Aid in resolving complex species groups with known cryptic diversity (e.g., Eurema, Spodoptera and Idaea).
- Inform environmental DNA (eDNA) based monitoring programs by providing comprehensive mitogenomic references [40].

Moreover, by integrating mitogenomics with emerging functional genomic tools like CRISPR, RNA-seq and WGS, researchers could explore questions of adaptive evolution, insecticide resistance and climate resilience in native species. For instance, comparing mitochondrial haplotype diversity across polluted versus pristine environments could reveal potential bioindicators among moth populations. The practical relevance is equally high in the domain of agricultural entomology, where accurate species identification of pest complexes is critical [41]. Groups like Noctuidae and

Crambidae, which include morphologically indistinguishable pests, stand to benefit enormously from mitogenome-based diagnostics.

Acknowledgements

The authors express their sincere gratitude to the Science and Engineering Research Board (SERB), Government of India, for their financial support of this research under the Core Research Grant (CRG) scheme (Sanction Order No. CRG/2020/001494, dated December 1, 2020).

Conflict of interest

No potential conflict of interest was reported by the author(s).

Author details

Aquib Majeed*, Tabraq Ali, Zahid Hussian and Sajad Hussain Parey
Insect Systematic Research Laboratory, Department of Zoology, School of Biosciences and Biotechnology, Baba Ghulam Shah Badshah University, Rajouri, India

*Address all correspondence to: aquibzoology@gmail.com

References

[1] Baral C, Baral HS, Inskipp C, Maharjan R. Lepidoptera diversity, richness, and distribution in semi-urban farmland and other habitats around Lumbini, Rupandehi. bioRxiv. 2025;**10**(13):2025-2001. DOI: 10.1101/2025.01.26.634957

[2] Thakur N, Bhardwaj V, Kumari P, Sidhu AK. An Inventory of Moths (Lepidoptera: Heterocera) from Kullu and Mandi Districts of Himachal Pradesh2024. DOI: 10.7324/JABB.2024.175339

[3] Chowdhury S, Dubey VK, Choudhury S, Das A, Jeengar D, Sujatha B, et al. Insects as bioindicator: A hidden gem for environmental monitoring. Frontiers in Environmental Science. 2023;**11**:1146052. DOI: 10.3389/fenvs.2023.1146052

[4] Kunte K, Basu DN, Kumar GG. Taxonomy, systematics, and biology of Indian butterflies in the 21st century. In: Indian Insects. CRC Press, Taylor and Francis; 2019. pp. 275-304. DOI: 10.1201/9780429061400

[5] Mutanen M, Wahlberg N, Kaila L. Comprehensive gene and taxon coverage elucidates radiation patterns in moths and butterflies. Proceedings of the Biological Sciences. 2010;**277**(1695):2839-2848. DOI: 10.1098/rspb.2010.0392

[6] Xing B, Lin L, Wu Q. Application of mitochondrial genomes to species identification and evolution. Electronic Journal of Biotechnology. 2025;**76**:39-48. DOI: 10.1016/j.ejbt.2025.04.001

[7] Antil S, Abraham JS, Sripoorna S, Maurya S, Dagar J, Makhija S, et al. DNA barcoding, an effective tool for species identification: A review. Molecular Biology Reports. 2023;**50**(1):761-775. DOI: 10.1007/s11033-022-08015-7

[8] Singh D, Kabiraj D, Sharma P, Chetia H, Mosahari PV, Neog K, et al. The mitochondrial genome of Muga silkworm (Antheraea assamensis) and its comparative analysis with other lepidopteran insects. PLoS One. 2017;**12**(11):e0188077. DOI: 10.1371/journal.pone.0188077

[9] Silva-Brandão KL, Lyra ML, Freitas AV. Barcoding Lepidoptera: Current situation and perspectives on the usefulness of a contentious technique. Neotropical Entomology. 2009;**38**:441-451. DOI: 10.1590/S1519-566X2009000400001

[10] Iwasaki M, Otaki JM. Synergistic Damage Response of the Double-Focus Eyespot in the Hindwing of the Peacock Pansy. Lepidoptera. 2017. pp. 65-78

[11] Jalali SK, Ojha R, Venkatesan T. DNA barcoding for identification of agriculturally important insects. In: New Horizons in Insect Science: Towards Sustainable Pest Management. 2015. pp. 13-23. DOI: 10.1007/978-81-322-2089-3_2

[12] Zheng C, Zhu X, Wang Y, Dong X, Yang R, Tang Z, et al. Mitogenomes provide insights into the species boundaries and phylogenetic relationships among three Dolycoris sloe bugs (Hemiptera: Pentatomidae) from China. Insects. 2024;**15**(2):134. DOI: 10.3390/insects15020134

[13] Dayrat B. Towards integrative taxonomy. Biological Journal of the Linnean Society. 2005;**85**(3):407-417. DOI: 10.1111/j.1095-8312.2005.00503.x

[14] Elameen A, Maduna SN, Mageroy MH, van Eerde A, Knudsen G, Hagen SB, et al. Novel insight into lepidopteran phylogenetics from the mitochondrial genome of the apple fruit moth of the family Argyresthiidae. BMC Genomics. 2024;**25**(1):21. DOI: 10.1186/s12864-023-09905-1

[15] Mitter C, Davis DR, Cummings MP. Phylogeny and evolution of lepidoptera. Annual Review of Entomology. 2017;**62**(1):265-283. DOI: 10.1146/annurev-ento-031616-035125

[16] Motyka M, Kusy D, Bocek M, Bilkova R, Bocak L. Phylogenomic and mitogenomic data can accelerate inventorying of tropical beetles during the current biodiversity crisis. eLife. 2021;**10**:e71895. DOI: 10.7554/eLife.71895

[17] Satam H, Joshi K, Mangrolia U, Waghoo S, Zaidi G, Rawool S, et al. Next-generation sequencing technology: Current trends and advancements. Biology. 2023;**12**(7):997. DOI: https://10.3390/biology12070997

[18] Yadav S, Sarangi S, Parmar S, Thodusu M, Soni S, Pandey K. Genetic engineering in insect management: New frontiers in pest control. Microbiology Research Journal International. 2024;**34**(12):106-121. DOI: 10.9734/mrji/2024/v34i121514

[19] Liu N, Fang L, Zhang Y. The complete mitochondrial genomes of four species in the subfamily Limenitidinae (Lepidoptera, Nymphalidae) and a phylogenetic analysis. Insects. 2021;**13**(1):16. DOI: 10.3390/insects13010016

[20] Joshi J, Agarwal I. Integrative taxonomy in the Indian subcontinent: Current progress and prospects. Journal of the Indian Institute of Science. 2021;**101**(2):125-149. DOI: 10.1007/s41745-021-00244-2

[21] Kalawate AS, Dinesh KP, Shabnam A. Morphological characterization and mt DNA barcode of a tiger moth species, Asota ficus (Fabricius, 1775) (Lepidoptera: Noctuoidea: Erebidae: Aganainae) from India. The Journal of Threatened Taxa. 2022;**14**(1):20503-20510. DOI: 10.11609/jott.7638.14.1.20503-20510

[22] Kim JS, Kim MJ, Jeong JS, Kim I. Complete mitochondrial genome of Saturnia jonasii (Lepidoptera: Saturniidae): Genomic comparisons and phylogenetic inference among Bombycoidea. Genomics. 2018;**110**(5):274-282. DOI: 10.1016/j.ygeno.2017.11.004

[23] Singh VK, Joshi PC, Gupta SK. Molecular and morphometric divergence of four species of butterflies (Nymphalidae and Pieridae) from the Western Himalaya, India. Molecular Biology Reports. 2020;**47**:8687-8699. DOI: 10.1007/s11033-020-05913-6

[24] Wiemers M, Keller A, Wolf M. ITS2 secondary structure improves phylogeny estimation in a radiation of blue butterflies of the subgenus Agrodiaetus (Lepidoptera: Lycaenidae: Polyommatus). BMC Evolutionary Biology. 2009;**9**:1-27. DOI: 10.1186/1471-2148-9-300

[25] Liu X, Qi M, Xu H, Wu Z, Hu L, Yang M, et al. Nine mitochondrial genomes of the Pyraloidea and their phylogenetic implications (Lepidoptera). Insects. 2021;**12**(11):1039. DOI: 10.3390/insects12111039

[26] Zou Y, Zhang Z, Zeng Y, Hu H, Hao Y, Huang S, et al. Common methods for phylogenetic tree construction and their implementation

in R. Bioengineering. 2024;**11**(5):480. DOI: 10.3390/bioengineering11050480

[27] Jacob SS, Sengupta PP, Chandu AGS, Shamshad S, Yogisharadhya R, Sudhagar S, et al. Existence of genetic lineages within Asian genotype of Taenia solium—Genetic characterization based on mitochondrial and ribosomal DNA markers. Transboundary and Emerging Diseases. 2022;**69**(4):2256-2265. DOI: 10.1111/tbed.14227

[28] Heaps SE, Nye TM, Boys RJ, Williams TA, Embley TM. Bayesian modelling of compositional heterogeneity in molecular phylogenetics. Statistical Applications in Genetics and Molecular Biology. 2014;**13**(5):589-609. DOI: 10.1515/sagmb-2013-0077

[29] Flouri T, Huang J, Jiao X, Kapli P, Rannala B, Yang Z. Bayesian phylogenetic inference using relaxed-clocks and the multispecies coalescent. Molecular Biology and Evolution. 2022;**39**(8):1-21. DOI: 10.1093/molbev/msac161

[30] Cornuault J, Sanmartín I. A road map for phylogenetic models of species trees. Molecular Phylogenetics and Evolution. 2022;**173**:107483. DOI: 10.1016/j.ympev.2022.107483

[31] Balakrishnan IK, Dubey H, Debnath R, Subrahmanyam G, Arunkumar PK, Pandey JP, et al. Complete characterization and comparative analysis of the mitochondrial genome of Indian tasar silkworm Antheraea mylitta (Lepidoptera: Saturniidae) and phylogenetic implications. The Journal of Asia-Pacific Biodiversity. 2025;**18**(1):108-117. DOI: 10.1016/j.japb.2024.07.006

[32] Zhu L, Han Y, Hou Y, Huang Z, Wang M, Chiba H, et al. Mitogenomic phylogenetic analyses provide novel insights into the taxonomic problems of several hesperiid taxa (Lepidoptera: Hesperiidae). Scientific Reports. 2023;**13**(1):7901. DOI: 10.1038/s41598-023-34608-8

[33] Thakur N, Sharma A, Kaur S, Ahluwalia KK, Sidhu AK, Kumar S, et al. Insect pest Spodoptera litura (Fabricius) and its resistance against the chemical insecticides: A review. Plant Science Today. 2024;**11**(4):192-203. DOI: 10.14719/pst.3078

[34] Ashfaq M, Akhtar S, Rafi MA, Mansoor S, Hebert PD. Mapping global biodiversity connections with DNA barcodes: Lepidoptera of Pakistan. PLoS One. 2017;**12**(3):e0174749. DOI: 10.1371/journal.pone.0174749

[35] do Nascimento BLS, da Silva FS, Nunes-Neto JP, de Almeida Medeiros DB, ACR C, da Silva SP, et al. First description of the mitogenome and phylogeny of Culicinae species from the Amazon region. Genes. 2021;**12**(12):1983. DOI: 10.3390/genes12121983

[36] Chen Q, Chen L, Liao CQ, Wang X, Wang M, Huang GH. Comparative mitochondrial genome analysis and phylogenetic relationship among lepidopteran species. Gene. 2022;**830**:146516. DOI: 10.1016/j.gene.2022.146516

[37] Pereira dos Santos M, Zotta Mota AP, Coiti Togawa R, Florencio Martins N, Bellard do Nascimento EF, Lucena VS, et al. The complete mitochondrial genome of Leucoptera coffeella (Lepidoptera: Lyonetiidae) and phylogenetic relationships within the Yponomeutoidea superfamily. Scientific Reports. 2024;**14**(1):7119. DOI: 10.1038/s41598-024-57869-3

[38] Venkataraman K, Sivaperuman C. Biodiversity hotspots in India.

In: Indian Hotspots: Vertebrate Faunal Diversity, Conservation and Management. Vol. 1. 2018. pp. 1-27. DOI: 10.1007/978-981-10-6605-4_1

[39] Ejigu GF, Jung J. Review on the computational genome annotation of sequences obtained by next-generation sequencing. Biology. 2020;**9**(9):295. DOI: 10.3390/biology9090295

[40] Sun S, Li S, Seim I, Du X, Yang X, Liu K, et al. Complete mitogenomes reveal high diversity and recent population dynamics in Antarctic krill. BMC Genomics. 2025;**26**(1):419. DOI: 10.1186/s12864-025-11579-w

[41] Breed MF, Harrison PA, Blyth C, Byrne M, Gaget V, Gellie NJ, et al. The potential of genomics for restoring ecosystems and biodiversity. Nature Reviews. Genetics. 2019;**20**(10):615-628. DOI: 10.1038/s41576-019-0152-0

Chapter 3

Seasonal and Habitat-Driven Specificity of Lepidopteran Host and Nectaring Plant Associations

Avisek Patra

Abstract

Seasons and habitats influence the selectivity with which lepidopteran insects choose their larval host plants and adult nectaring options. The host and nectar plant relationships of various lepidopteran species in semi-urban, grassland, forest edge, and agricultural environments were investigated in this study. Field observations showed that habitat layout and seasonal floral availability had a significant impact on plant choices. While some species adapted opportunistically, especially during times of resource scarcity, others maintained tight host plant fidelity. The ecological significance of habitat integrity and plant diversity in sustaining lepidopteran populations is emphasized by the study. These results provide significant insights for managing habitat and conserving pollinators in dynamic environments.

Keywords: lepidoptera, host plant specificity, nectaring plant, seasonal variation, plant-insect interaction, biodiversity conservation

1. Introduction

Plant-insect interactions serve as the structural and functional framework for terrestrial ecosystems, sustaining essential ecological processes such as pollination, herbivory, and nutrient cycling. These interactions tend to be highly specialized, formed by coevolutionary relationships that have linked insect taxa to specific plant partners for millions of years [1, 2]. Lepidoptera, an insect order that includes butterflies and moths, is one of the most ecologically significant and diverse, with over 180,000 known species [3]. Lepidopterans have two ecological roles: herbivores in their larval stages and pollinators in their adult form [4, 5].

Butterflies, an attractive and thoroughly studied subgroup of Lepidoptera, are well recognized as bioindicators due to being sensitive to habitat degradation, climate variability, and changes in land use [6]. Their lifecycles are strongly related to plants, however, via distinctive ecological pathways. Larvae are host plant specific, often limited to one or a few plant species, as determined by egg-laying preferences and evolved through chemical compatibility [7]. As adults, butterflies transition from herbivory to nectar consuming, relying on floral resources not just for survival but also to maintain important behaviors such as mating and dispersal [4].

These host-nectar plant interactions are more than just behavioral advantages; they indicate fundamental coevolutionary links. Floral shape and nectar content, for example, are frequently fine-tuned to meet Lepidopteran proboscis length, feeding habit, and sensory capacities [8, 9]. This ecological relationship promotes both plant reproduction and insect longevity, increasing the sustainability of mutually beneficial networks.

These associations' dynamic structure, however, is quite susceptible to variations in environment and season. Temperature, photoperiod, and rainfall variations throughout the year affect the phenology of plants and the availability of floral resources, which in turn affects insect activity and behavior [10, 11]. Similarly, different habitat types provide variable levels of resource continuity and plant diversity, including semi-urban landscapes, grasslands, forest edges, and agricultural margins [12, 13].

Lepidoptera are perfect models for studying the specificity and adaptability of plant use throughout developmental stages because of their dual ecological roles. The degree of consistency varies among species; some remain strictly specialized on specific hosts or nectar plants, while others utilize opportunistic strategies, particularly when habitat is disturbed or flowers are unavailable [14, 15]. The stability of ecosystems and population stability are greatly influenced by these variations.

Furthermore, in view of global environmental change, it is crucial to understand whether species are specialized or generalists and how this influences their ecological stability. While generalists may adapt more easily but provide distinct contributions to ecological networks, specialists are typically more susceptible to habitat fragmentation and phenological mismatches [16].

The purpose of this study is to examine how lepidopteran host and nectaring plant interactions change ecologically and in behavior across different habitat types and seasonal variations. It aims to (i) analyze patterns of larval host and adult nectar plant selection; (ii) study the impact of habitat heterogeneity and seasonal floral change on foraging behavior; and (iii) evaluate the implications for conservation strategies aimed at maintaining lepidopteran diversity and ecosystem function in dynamic environments (**Figure 1**).

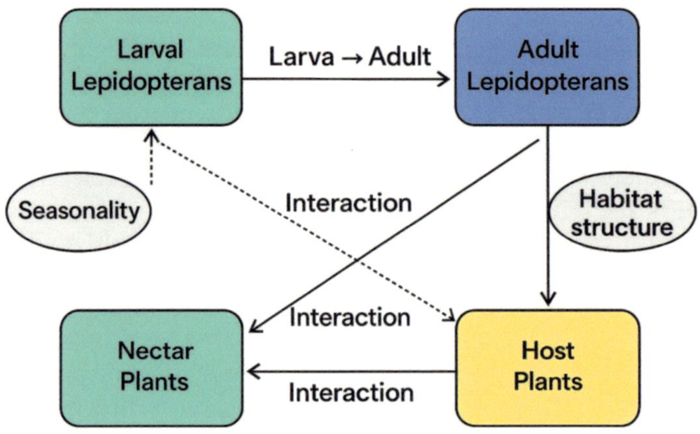

Figure 1.
Conceptual framework of lepidopteran plant associations across life stages and ecological drivers.

2. Theoretical overview and conceptual framework

An integrated viewpoint based on behavioral ecology, coevolutionary theory, and landscape-level environmental variation is necessary to comprehend the ecological connections between Lepidoptera and their host and nectaring plants. The theoretical frameworks that underlie the ecological specialization and behavioral strategies of moths and butterflies throughout life stages and conditions in nature are described in this section.

2.1 Host plant specificity in lepidoptera

Host plant specificity is an important characteristic in lepidopteran ecology. It refers to the level to which larvae depend on a small number of plant species for feeding and development. Many lepidopteran species exhibit monophagy or oligophagy, feeding on one or a few closely related host species [17]. This specialization is primarily mediated by female egg-laying behavior, which is influenced by plant secondary metabolites, surface chemistry, and visual or olfactory cues [7].

From an evolutionary perspective, coevolution—in which insects and plants apply selective pressures to one another—is frequently the source of such specialization. The close relationship between milkweeds (*Asclepias* spp.) and monarch butterflies (*Danaus plexippus*) is a classic example, where plant toxin tolerance and larval feeding behavior have coevolved [18]. Specialization can make larvae more susceptible to habitat loss and host plant shortage, even when it improves their performance on ideal hosts [19].

Pyrolizidine alkaloids, which are essential for both larval development and adult pheromone synthesis, are found in plants of the Solanaceae family, to which the glass-wing butterfly (*Greta oto*) shows integrity. Though it has been seen using exotic hosts like *Plantago lanceolata* in disturbed habitats, the Baltimore checkerspot (*Euphydryas phaeton*) is primarily associated with turtlehead (*Chelone glabra*). This behavior has fitness trade-offs and raises concerns about ecological traps [14].

Phylogenetically, host specificity can also be conserved. According to studies, a large number of Lycaenid butterflies associate with specific host plant groups, like the Fabaceae and Myrtaceae, frequently through mutualism mediated by ants. These coevolved connections suggest that host specialization may influence species diversification and niche adaption over macroevolutionary periods, suggesting that it may be more than just an ecological convenience [20].

Specialization has an ecological cost even while it increases larval success and lessens interspecific competition. When host plant supply is disrupted by habitat fragmentation or phenological mismatches, species with strong host specificity may be significantly affected.

2.2 Nectaring plant use in adult butterflies

While larval host specialization in Lepidoptera is widely known, adult butterflies' nectaring behavior is a complementary and equally important ecological interaction. Adult Lepidoptera depend on floral nectar for their basic energy supply, especially during flight, mate-seeking, dispersal, and reproduction. Nectar composition, volume, accessibility, and geographical distribution influence both butterfly behavior and plant reproductive performance, resulting in a mutualistic and dynamic plant-pollinator relationship [4, 5].

In general, adult butterflies are considered to be more generalists regarding when it comes to floral preferences than larvae with host plants. However, this generality changes significantly depending on species, environmental conditions, and resource availability. The painted lady (*Vanessa cardui*), for example, has diverse nectaring preferences and visits over 100 plant species during its long migrations, including thistles (*Cirsium* spp.), milkweed (*Asclepias* spp.), and alfalfa (*Medicago sativa*). This broad range of nutrition allows it to adapt to a variety of environments and floral groups on a continental scale.

In contrast, some species have definite floral preferences that correspond to their morphological characteristics. The pipevine swallowtail (*Battus philenor*), with a long and narrow proboscis, prefers tubular flowers like honeysuckle (*Lonicera* spp.) and *Penstemon* species, which offer more nectar on every visit. Heliconiine butterflies, such as *Heliconius erato*, have coevolved with passionflowers (*Passiflora* spp.), from which they take nectar as well as pollen—an unusual characteristic among Lepidoptera that improves nitrogen intake and increases adult lifespan [21].

Some butterflies also exhibit temporal partitioning of nectar consumption, adjusting to seasonal flower supply. In tropical gardens, the common Mormon (*Papilio polytes*) changes between *Ixora*, *Lantana*, and *Clerodendrum* species based on their phenological stages. This behavioral adaptability maintains a consistent energy supply throughout the flying duration and promotes reproductive output.

Nectar chemistry is important in addition to floral features. Certain butterflies prefer sucrose-rich nectar over glucose or fructose, which can affect foraging efficiency and visiting rates. The silver-washed fritillary (*Argynnis paphia*) prefers to eat on bramble (*Rubus fruticosus*) blooms due to their high nectar volume and sugar concentration [22].

Furthermore, in disturbed or resource-limited habitats, some species demonstrate adaptive flexibility. The gray pansy (*Junonia atlites*), which is typically found in open landscapes, readily feeds on a diverse range of ruderal flowers such as *Tridax procumbens* and *Chromolaena odorata*, demonstrating how generalists can survive in urban or degraded habitats.

Therefore, an important functional relationship between plant phenology, habitat design, and pollinator performance is represented by nectaring behavior in adult butterflies. Knowing these relationships helps us better understand pollination network structure and ecological resilience, especially in the context of environmental stressors like habitat loss or climate-driven floral alterations.

2.3 Specialist vs. generalist strategies in plant use

Lepidopteran species use resources in a variety of ways, from strict experts that only eat or drink from a small group of plants to broad generalists that eat a lot of different plant types. These tactics are the result of evolutionary choices influenced by habitat stability, physiological limitations, and ecological stresses. It is essential to understand these adaptive patterns in order to forecast how species will react to changes in their environment and to create focused conservation plans.

Specialist species often establish close associations with a small set of plants, a trend found particularly in larvae. These animals typically have specially developed sensory systems, detoxification routes, and behavioral adaptations that are exclusive to their hosts. For example, the Baltimore checkerspot (*Euphydryas phaeton*) has long been connected with turtlehead (*Chelone glabra*), and its reproductive success is closely related to the phenology and wellness of this one plant species. Although it can occasionally utilize new hosts such as *Plantago lanceolata*, these changes may jeopardize larval survival and adult fecundity [14].

The Papilionidae family butterfly *Parides iphidamas* exhibits strong loyalty to *Aristolochia* plants, which contain harmful aristolochic acids, in tropical forests. In order to protect themselves from predators, the larvae not only tolerate these substances but also sequester them. These butterflies' limited host range makes them susceptible to habitat deterioration or host plant loss, even if it has been advantageous for evolution in stable conditions.

Although they are less prevalent, nectar specialists do exist at the adult stage. Specialized pollen and nectar-feeding habits are displayed by some *Heliconius* butterfly species on particular *Passiflora* species, which also act as larval hosts. Although it increases dependence on intact, biodiverse ecosystems, this dual-stage specialization strengthens a profound coevolutionary link [21].

Generalist species, on the other hand, can make use of a wide variety of plants. A well-known example of a generalist herbivore, the cosmopolitan cabbage white (*Pieris rapae*) consumes a wide variety of weedy and cruciferous crops. Its adaptable oviposition behavior and tolerance for varying host quality are the main reasons for its success in disturbed agricultural settings [17].

The painted lady (*Vanessa cardui*), another generalist, exhibits remarkable migratory adaptation by nectaring from a broad range of flowering species across continents and feeding on over 300 larval host plants. It can take advantage of a variety of climate zones and seasonal floral pulses thanks to its ecological flexibility.

Another amazing example is the common buckeye (*Junonia coenia*). While its adults nectar from generalist plants like *Lantana camara* and *Tridax procumbens*, its larvae feed on a variety of host species, such as *Plantago*, *Antirrhinum*, and *Ruellia*. In environments that have been fragmented or altered by humans, this dual-stage generalism enables increased resilience.

High efficiency and host-plant matching are advantageous to specialists, and they frequently result in improved larval performance and stronger coevolutionary relationships. They are more vulnerable to environmental changes, though, including phenological mismatches or habitat fragmentation. Conversely, generalists have adaptive advantages in unstable or damaged ecosystems but may perform worse on any one host [23].

Community dynamics are also impacted by these trade-offs. In species interactions, specialist-rich communities are typically more stable, but they are also more susceptible to abrupt changes. Although generalist-dominated assemblages may be more robust, they may be less effective at trophic specialization or pollination.

2.4 Seasonality as an ecological filter

One of the most important abiotic filters influencing the temporal dynamics of plant–insect interactions is variations in the seasons, which is controlled by changes in temperature, precipitation, and photoperiod. Seasonal fluctuation determines the availability of larval host plants and adult nectaring supplies for Lepidoptera, whose life cycles are strongly linked to plant phenology and environmental cues. With serious ecological repercussions, these variations may synchronize or misalign life-history events such as egg laying, larval development, emergence, and feeding [10, 24].

2.4.1 Phenological synchrony and mismatch

Phenological synchrony, or the matching of insect development stages with the availability of hosts or nectar plants, is essential for the persistence of species in temperate environments. For example, early spring crucifers like *Cardamine pratensis* are essential to the orange tip butterfly (*Anthocharis cardamines*). Larvae may lose the best eating window if emergence is postponed because of cooler springs, which would lower their fitness and survival.

However, phenological mismatches, which are becoming more and more common in climate change situations, can separate insect emergence from foliar or floral peaks. For specialists, this phenomenon is very harmful. For instance, blackthorn (*Prunus spinosa*) branches are necessary for the brown hairstreak (*Thecla betulae*) to lay eggs; early budding brought on by warmer winters may cause the eggs to desiccate before hatching [25].

Even while variations in seasons are frequently less noticeable in tropical systems, resource restrictions are nonetheless imposed rhythmically by dry and wet seasons. During both wet and dry seasons, butterflies like the common crow (*Euploea core*) move their nectar-gathering habits across species like *Rauvolfia*, *Lantana*, and *Clerodendrum*. Although it differs among species and habitat types, this behavioral flexibility improves survival.

2.4.2 Seasonal adaptations in life history

Different tactics are used by lepidopteran species to deal with seasonal uncertainty. Multiple generations are produced annually by multivoltine species, such as *Pieris brassicae*, which enables quick exploitation of fluctuating floral resources. Univoltine species, like *Parnassius apollo*, on the other hand, are more closely synced to high-altitude floral windows, which leaves them susceptible to even a small amount of phenological drift.

Furthermore, migratory and dormancy behavior are also influenced by seasonal restrictions. While certain butterflies, like the comma butterfly (*Polygonia c-album*), enter reproductive diapause to compensate for floral deficiencies, migratory butterflies, like *Danaus plexippus*, use long-distance travel to avoid winter scarcity.

2.4.3 Implications for resource use and habitat matching

Seasonal variation interacts with habitat structure rather than acting independently. Non-native ornamental plants can temporarily fill in seasonal nectar deficits, resulting in longer blooming periods in urban and fragmented settings. Such manufactured continuity, however, could potentially create biological traps by tricking ovipositing females into depositing eggs on inappropriate or undernourished hosts.

These relationships emphasize how important it is for butterflies and their plant partners to have the same phenology. Understanding the mechanisms of temporal filtering is crucial for the conservation of phenology-sensitive lepidopteran species and the maintenance of stable plant-insect interaction networks as climate-driven changes continue to modify seasonal rhythms.

2.5 Habitat heterogeneity and plant resource distribution

Lepidopteran utilization of resources and population persistence are significantly influenced by habitat heterogeneity, which is the regional variety in environmental conditions, plant composition, and structural features. A variety of microhabitats found in diverse landscapes affect the host and nectar plants' temporal and geographical availability. On the other hand, settings that are overly simple or fragmented might break important ecological connections, particularly for species that have specific ecological needs [26, 27].

2.5.1 Influence of habitat structure on resource access

The diversity and abundance of native flowering plants in natural grasslands and forest edge environments produce a consistent and reliable foraging landscape. Butterflies such as the chalkhill blue (*Polyommatus coridon*) flourish in European calcareous grasslands because of their high host plant fidelity to *Hippocrepis comosa*, a leguminous herb that predominates in these habitats [28]. Additionally, these habitats provide a variety of nectar supplies that let adult populations survive for prolonged durations of flight.

On the other hand, the spatial isolation of plant patches in fragmented or urbanized landscapes may restrict the number of oviposition locations and foraging opportunities. Butterflies like the lemon pansy (*Junonia lemonias*) and little white (*Pieris rapae*) may survive in urban green spaces because of their generalist diets, although they frequently depend on invasive or ornamental nectar sources like *Tridax procumbens* and *Lantana camara* [12]. Reproductive fitness may be impacted by these substitute resources' probable lack of native species' phenological match or nutritional balance.

2.5.2 Agricultural land-use and floral resource depletion

Lepidopteran populations face a special challenge as a result of agricultural intensification. Monocultures substitute single-species crops, which frequently lack the structural or chemical characteristics necessary for host or nectar compatibility, with varied floral assemblages. For instance, research in Mediterranean agroecosystems has revealed a dramatic drop in butterfly diversity associated with the loss of hedgerows and fallow fields, which historically functioned as nectar corridors and host plant refugia [29].

However, hedgerows, buffer strips, and field margins can serve as important dispersal stepping stones and floral variety repositories. These semi-natural features have been widely used by butterflies such as the meadow brown (*Maniola jurtina*) and common blue (*Polyommatus icarus*) for mating and nectaring activities [30].

2.5.3 Landscape connectivity and metapopulation dynamics

The degree of landscape connectivity, or how easily butterflies may migrate between resource patches, is crucial for population stability in addition to patch composition. While sedentary or low-dispersal species like *Melitaea cinxia* rely on spatially coherent habitats for metapopulation persistence, highly mobile species like *Vanessa cardui* and *Colias croceus* can successfully cross fragmented habitats [30]. Pollination network failure, local extinctions, and genetic isolation can result from poor connection.

Furthermore, plant-insect interactions are further modulated by microclimatic variation among habitats. For example, butterflies like *Argynnis paphia* take advantage of the thermal gradients and wind-sheltered foraging places found around forest edges, particularly during periods of high midday activity when open spaces may become thermally demanding.

The structural makeup and spatial layout of habitats, in addition to the existence of certain plant species, determine the ecological integrity of lepidopteran–plant interactions. It is more likely to support a variety of specialists and generalists when habitat mosaics consisting of native grasslands, forest margins, hedgerows, and lightly disturbed greenspaces are maintained. Therefore, in order to maintain the full life-cycle requirements of various lepidopteran species, conservation initiatives should prioritize both habitat continuity and resource variety.

3. Description of study systems and ecological conditions

Understanding the ecological circumstances in which lepidopterans interact with their hosts and nectar plants is critical for understanding their behavioral distinctiveness and adaptability. The study examined butterfly-plant relationships in four habitat types: semi-urban greenspaces, natural grasslands, forest edges, and agricultural edges. These habitats vary in plant diversity, human disturbance, and resource predictability [10, 29, 31].

3.1 Semi-urban greenspaces

Semi-urban greenspaces, including green corridors, institutional campuses, urban gardens, and municipal parks, represent human-modified habitats that often support a mosaic of ornamental and native vegetation. These environments are typically characterized by continuous floral availability but limited larval host diversity. Commonly cultivated flowering plants such as *Catharanthus roseus*, *Lantana camara*, *Ixora coccinea*, *Bougainvillea glabra*, and *Tagetes erecta* serve as consistent nectar sources for butterflies throughout the year, making these habitats suitable for generalist species. Butterflies such as *Danaus chrysippus*, *Junonia lemonias*, *Catopsilia pomona*, and *Eurema hecabe* are frequently observed here due to their broad ecological tolerance and ability to exploit a wide range of nectar sources and occasional ornamental host plants.

However, these habitats face constant anthropogenic pressures including pesticide use, pruning, artificial lighting, and construction activities, which influence microclimatic conditions and plant-insect interactions. While such disturbances limit the presence of habitat specialists, they often favor ecologically flexible species capable of adjusting their foraging, reproductive, and phenological strategies. Furthermore, these greenspaces play an important role in urban conservation planning by acting as stepping-stone habitats that enhance landscape connectivity and provide platforms for environmental education and citizen science initiatives [10, 29, 32, 33].

3.2 Natural grasslands

Natural grasslands are open-canopy ecosystems dominated by native grasses, herbaceous plants, and a rich diversity of seasonal wildflowers. These habitats are shaped by climatic rhythms, particularly the monsoonal cycle, which governs vegetative growth and floral succession. After the monsoon rains, a surge in herbaceous flora such as *Crotalaria* spp., *Tridax procumbens*, *Alysicarpus* spp., *Ageratum conyzoides*, and *Corchorus* spp. offers a critical window of abundance for both larval and adult butterfly stages. These plant communities support a wide range of butterfly species, with particular importance for Lycaenids like *Zizeeria karsandra* and *Chilades pandava*, which exhibit narrow host preferences and are highly dependent on native Fabaceae and other legumes.

In addition to Lycaenids, grasslands also harbor Nymphalid butterflies like *Ypthima asterope* and *Junonia hierta*, which utilize low-lying vegetation for both oviposition and adult foraging. Pierids such as *Eurema brigitta* and *Pieris brassicae* are also frequently recorded due to the availability of host plants like *Capparis* and *Brassica*-related weeds in these open habitats.

Despite their ecological richness, natural grasslands face escalating threats from land-use changes, including conversion to agriculture, urban encroachment, overgrazing, and fire mismanagement. Fragmentation not only reduces the extent of available habitat but also isolates populations, thereby threatening species with low dispersal ability or high host specialization [34, 35]. Nevertheless, when preserved, grasslands serve as important biodiversity hotspots that sustain complex plant-insect networks and support ecosystem resilience.

Moreover, grasslands offer microclimatic buffering and structural heterogeneity that are vital for thermoregulation, oviposition site selection, and predator avoidance in butterflies. Recent observations have shown that seasonally dry grasslands in central and western India support significant butterfly richness, with over 50 species recorded during the post-monsoon period alone [33]. Their importance is especially pronounced in semi-arid landscapes where forest cover is sparse, making them critical for conservation-focused management and pollinator-friendly land-use planning.

3.3 Forest edge habitats

Forest edge habitats, also known as ecotones, represent transitional zones where closed-canopy forests meet open landscapes such as agricultural lands or grasslands. These zones offer structurally complex environments that exhibit a gradient of light, moisture, and plant diversity, making them particularly attractive to a wide range of butterfly species. Such areas provide a combination of shaded understory vegetation and sunlit herbaceous layers, supporting both larval development and adult foraging needs.

The flora typically found in forest edge habitats includes species like *Clerodendrum infortunatum*, *Stachytarpheta indica*, *Rauvolfia serpentina*, and the commonly invasive yet nectar-rich *Lantana camara*, which serve as host or nectar plants for butterflies such as *Hypolimnas bolina*, *Euploea core*, and *Parantica aglea* [10, 36]. These plant-butterfly relationships underscore the functional significance of ecotones as breeding and feeding grounds, especially for shade-loving species and those with large body sizes that require thermally buffered conditions.

Butterflies like *Tanaecia lepidea* and *Euthalia aconthea*, which are sensitive to thermal extremes, are frequently found in forest margins where canopy shade helps regulate their body temperature [31]. These areas also act as ecological corridors that enable gene flow between isolated forest patches, thus playing a critical role in species persistence in fragmented landscapes [29].

Seasonal patterns further enhance the ecological value of forest edges. During the monsoon and early post-monsoon periods, these zones experience a flush of flowering in both ground and shrub layers. This attracts a range of butterfly species such as *Papilio polytes*, *Neptis hylas*, *Ypthima baldus*, and *Mycalesis perseus*, which are observed using these habitats extensively for nectar foraging and courtship behaviors [32, 37].

Importantly, the dual availability of host and nectar plants makes forest edges highly productive for butterflies with complex life cycles. The presence of vines and layered vegetation also supports vertical habitat use by different species, further enhancing biodiversity. However, these areas are vulnerable to anthropogenic activities such as illegal logging, grazing, and the spread of invasive plant species like *Chromolaena odorata* and *Mikania micrantha*, which can outcompete native flora and disrupt critical plant-insect interactions [30, 38].

Despite these threats, forest edges have shown to support an overlapping mix of generalist and forest-specialist species, often exhibiting higher butterfly richness than the forest interior or adjacent agricultural lands [11]. Their role as buffer zones and refuges makes them essential in conservation strategies that aim to maintain ecological connectivity and mitigate the effects of habitat fragmentation.

3.4 Agricultural margins and fallow lands

Agricultural margins and fallow lands represent dynamic and ecologically opportunistic habitats often overlooked in butterfly conservation. These zones, typically characterized by bunds, hedgerows, weedy patches, and uncultivated plots between crop fields, serve as vital ecological refuges, particularly in highly modified rural landscapes. Such areas often harbor diverse weed communities and spontaneous herbaceous flora that flourish in the absence of intensive tillage or herbicide application, providing a continual though irregular supply of floral resources across seasons.

Common nectar sources observed in these environments include *Tridax procumbens*, *Cleome viscosa*, *Ipomoea cairica*, *Vernonia cinerea*, and *Leucas aspera*, which attract nectar-feeding adult butterflies throughout the year [29, 32]. These flowering species are especially abundant after monsoon rains and contribute significantly to adult butterfly nutrition during pre- and post-monsoon periods.

Larval host plants in these habitats often consist of fast-growing ruderal species such as *Cassia tora*, *Amaranthus viridis*, *Portulaca oleracea*, *Chrozophora rottleri*, and *Boerhavia diffusa*, all of which support herbivorous lepidopteran larvae [10, 34]. These support generalist species like *Catopsilia pomona*, *Pieris brassicae*, *Eurema hecabe*, and *Spodoptera litura*, which are capable of exploiting a wide range of resources due to their ecological flexibility and polyphagous habits.

Although resource-rich, these areas are subject to fluctuating levels of disturbance. Practices such as herbicide spraying, plowing, and monoculture planting of crops like rice, wheat, or cotton often lead to temporal instability in floral availability. Frequent pesticide use not only reduces butterfly diversity but also selects against specialist species with narrow dietary preferences, such as *Colotis etrida* and *Danaus genutia*, which fail to find suitable larval hosts or safe foraging environments in such altered settings (**Table 1**) [24, 26].

Despite these limitations, agricultural margins can serve as temporary stepping stones for butterfly dispersal and act as seasonal breeding grounds when less-disturbed conditions prevail. Studies from semi-arid regions in central and southern India have shown that butterfly diversity can remain relatively high in fallow fields, especially during the post-monsoon period, due to the proliferation of both larval and nectar plants [37, 39].

In particular, butterflies like *Junonia orithya*, *Eurema brigitta*, and *Danaus chrysippus* have been observed to lay eggs on weedy margins, where host plants such as *Ricinus communis* and *Calotropis procera* occur sporadically [12, 36]. Additionally,

Species	Host Plant(s)	Nectar Plant(s)	Habitat	Seasonality	Specialization
Danaus chrysippus	*Calotropis spp.*	*Lantana camara*	Urban gardens	Year-round	Specialist
Papilio polytes	*Citrus spp.*	*Ixora spp.*	Forest edges	Post-monsoon	Generalist
Eurema hecabe	*Cassia spp.*	*Tridax procumbens*	Grasslands	Monsoon	Generalist
Catopsilia pomona	*Senna spp.*	*Clerodendrum spp.*	Agricultural fields	Winter	Specialist
Junonia atlites	*Plantago spp.*	*Ageratum spp.*	Semi-urban areas	Summer	Generalist
Pieris brassicae	*Brassica spp.*	*Tagetes spp.*	Urban-rural fringe	Spring	Specialist
Graphium agamemnon	*Annona spp.*	*Bauhinia spp.*	Woodlands	Year-round	Generalist
Tirumala limniace	*Asclepias spp.*	*Jasminum spp.*	Scrublands	Monsoon	Specialist
Hypolimnas bolina	*Portulaca spp.*	*Hibiscus spp.*	Parks	Pre-monsoon	Opportunist
Delias eucharis	*Capparis spp.*	*Tecoma stans*	Hill slopes	Late winter	Specialist
Acraea violae	*Passiflora spp.*	*Cosmos spp.*	Meadows	Post-monsoon	Generalist
Neptis hylas	*Mussaenda spp.*	*Ruellia spp.*	Riverbanks	Summer	Generalist
Lethe rohria	*Bambusa spp.*	*Impatiens spp.*	Bamboo groves	Spring	Specialist
Ypthima asterope	*Grasses*	*Bidens pilosa*	Open woodlands	Year-round	Generalist
Cepora nerissa	*Crataeva spp.*	*Heliotropium spp.*	Riparian zones	Winter	Specialist
Appias albina	*Drypetes spp.*	*Duranta spp.*	Secondary forests	Monsoon	Specialist
Zizina otis	*Clover spp.*	*Vernonia spp.*	Roadside herbs	Spring	Opportunist
Tarucus indica	*Zizyphus spp.*	*Euphorbia spp.*	Shrublands	Summer	Specialist
Catochrysops strabo	*Albizia spp.*	*Leucas spp.*	Savannah	Post-monsoon	Generalist
Pseudergolis wedah	*Strobilanthes spp.*	*Chromolaena odorata*	Monsoon forests	Year-round	Specialist

Table 1.
Some lepidopteran species with corresponding host and nectar plants across diverse habitats and seasons.

small puddles and muddy areas formed along irrigation bunds in such landscapes provide essential sources of salts and nutrients for male butterflies engaging in puddling behavior.

To enhance their ecological value, these habitats can be integrated into butterfly conservation strategies by maintaining uncultivated strips, reducing pesticide use near bunds, and encouraging native weed growth during non-crop seasons [32, 35]. Their proximity to both human settlements and natural patches makes them critical in maintaining landscape connectivity and promoting metapopulation resilience in butterfly communities (**Figure 2**) [28, 40].

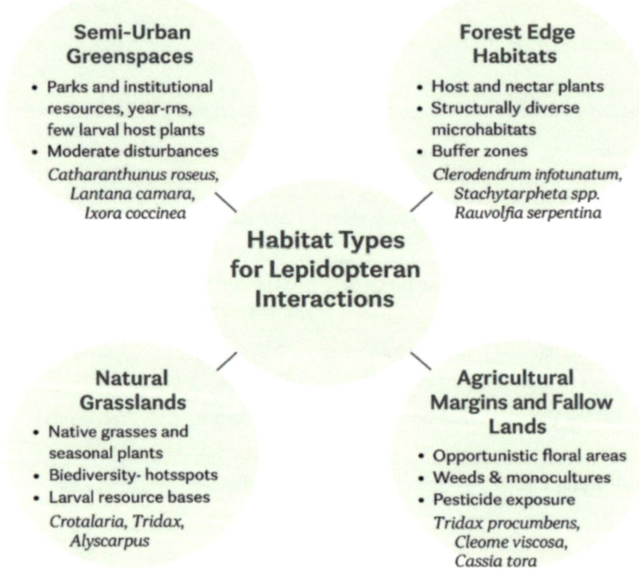

Figure 2.
Ecological landscape mosaic and habitat utilization patterns of lepidoptera.

3.5 Climatic and seasonal context

All habitats are part of a subtropical monsoonal climate system with distinct seasonal transitions that affect plant and insect phenology. The region experiences three distinct seasons: pre-monsoon (March–May), monsoon (June–September), and post-monsoon (October–January). During the pre-monsoon season, arid conditions prevail with few floral resources, limiting both larval and adult nutrition. The monsoon season stimulates vegetative growth, increasing the availability of both host plants for oviposition and a wide variety of floral nectar sources. This is the best time for plant–insect interactions, with many lepidopteran species emerging simultaneously with flower abundance.

In a study from the Bankura district of West Bengal, the monsoon season recorded the highest butterfly abundance among all seasons, underscoring the climatic influence on lepidopteran activity and habitat use [41]. The post-monsoon season has moderate humidity and temperature, allowing for extended feeding and reproductive activity before a gradual drop in bloom. These seasonal changes have a significant impact on butterfly life cycles, including voltinism, host fidelity, and foraging behaviors. Furthermore, mismatches between flowering phenology and butterfly emergence, which are frequently caused by climate variability, can alter resource availability and ecological fitness [10, 24].

Recognizing these temporal fluctuations is critical for understanding how lepidopteran–plant interactions persist or change across ecological gradients.

4. Patterns of host and nectar plant use across ecological contexts

Butterflies' selection of larval hosts and adult nectar plants demonstrates impressive ecological and evolutionary adaptations. These decisions are not made at random;

rather, they are influenced by a complex interaction between habitat structure, seasonal dynamics, and species-specific oviposition and foraging tactics. This section looks at how lepidopteran plant relationships differ in various ecological environments by combining field observations and published data.

4.1 Seasonal patterns in plant use

Butterfly–plant interactions are temporally dynamic and mostly determined by climatic rhythms. In tropical and subtropical regions, seasonality is one of the most important factors of resource availability. Studies in Peninsular India, Bangladesh, and Southeast Asia consistently show that the monsoon and post-monsoon seasons have the highest butterfly richness and activity [10, 41, 42]. During these times, blooming species such as *Tridax procumbens*, *Vernonia cinerea*, and *Ageratum conyzoides* provide the most nectar, while host plants such as *Cassia tora*, *Crotalaria juncea*, and *Portulaca oleracea* provide rich, nutritious food for larvae.

During the pre-monsoon season, high temperatures and dry conditions limit flower diversity. As a result, butterfly diversity plummets, with only a few drought-tolerant species such as *Danaus chrysippus* and *Catopsilia pomona* remaining active, relying on perennial floral species such as *Lantana camara* and *Bougainvillea spectabilis* [32, 37, 43].

Junonia orithya and *Eurema brigitta* have been observed in urban settings using ornamental plants such as *Catharanthus roseus* and *Ixora coccinea*, which flower all year and supply constant nectar [41]. This seasonal resource buffering is frequently unavailable in natural systems, as butterfly emergence must coincide with host plant sprouting and flower bloom times.

Furthermore, phenological mismatches induced by climate change are becoming more apparent. For example, research in the Western Ghats has found an asynchronous relationship between butterfly emergence and nectar resource availability in species such as *Graphium agamemnon* and *Papilio helenus* [39]. Similar patterns have been observed in temperate regions for *Gonepteryx rhamni* [25], highlighting the global importance of seasonal plant–insect interaction.

4.2 Habitat-driven variation in host and nectar plant associations

Butterfly resource usage varies dramatically among habitat types, and ecological composition influences both the availability and diversity of larval and adult resources. Natural grasslands with open-canopy plants, native legumes, and wildflowers make perfect homes for specialist butterflies. In such grasslands, species including *Zizeeria karsandra*, *Tarucus nara*, and *Leptotes plinius* are frequently found with *Indigofera*, *Alysicarpus*, and *Desmodium* spp. [36, 37]. Nectar plants such as *Blumea lacera* and *Vernonia cinerea* provide additional assistance during the adult stages.

In contrast, forest edge habitats have a distinct mix of shade-tolerant plants and vertical stratification of vegetation, which supports both nectar and host plant specialists. Butterflies like *Hypolimnas bolina*, *Tanaecia lepidea*, and *Euploea core* are commonly observed using host plants such as *Ficus hispida*, *Nerium oleander*, and *Passiflora foetida* and forage on *Clerodendrum infortunatum*, *Stachytarpheta jamaicensis*, and *Lagerstroemia speciosa* [29, 35].

Semi-urban greenspaces, such as parks and institutional gardens, are dominated by beautiful and exotic plants. While these locations frequently lack native larval hosts, they do provide constant nectar sources. *Papilio demoleus* and *Junonia lemonias* use plants including *Catharanthus roseus*, *Hibiscus rosa-sinensis*, and *Duranta erecta* for nectar. *Citrus* spp., *Cassia occidentalis*, and *Calotropis gigantea* are patchy host plants that are frequently maintained on roadside verges.

Agricultural landscapes, particularly fallow fields and bunds, include a unique mix of ruderal and weedy plants. Butterflies including *Pieris canidia*, *Catopsilia pyranthe*, and *Eurema hecabe* feed on crop hosts such as *Brassica*, *Senna*, and *Cleome viscosa*. *Tridax procumbens*, *Ipomoea carnea*, and *Corchorus* are the most common nectar sources around field boundaries [13, 38]. However, pesticide drift and monoculture techniques lead to temporal constraints in both larval and adult resources.

4.3 Host plant specialization: Constraints and adaptive value

Butterfly host plant specialization is influenced by ecological stability, plant chemistry, and evolutionary history. *Zemeros flegyas* is closely related to *Maesa indica*, whereas *Ariadne merione* only consumes *Trema orientalis*. Coevolved plant–insect chemical cues are frequently involved in these interactions, which facilitate optimal larval development and predator avoidance [14, 20].

This intense specialization has drawbacks, though. Specialist populations in fragmented habitats are directly threatened by the loss or reduction of host plants. Due to the loss of host trees in deforested corridors, *Tanaecia lepidea* and *Appias albina* are locally extinct, according to studies conducted in Sri Lanka and Western India [34, 37].

Generalist species, on the other hand, such as *Catopsilia pomona*, *Eurema blanda*, and *Pieris brassicae*, show great flexibility and larval tolerance across a variety of plant groups. Because of their adaptability, they may survive in altered habitats and take advantage of plants that are ephemeral or linked with humans, such as *Portulaca oleracea*, *Senna alata*, and *Brassica juncea* [7, 25].

4.4 Nectaring behavior: Preference, plasticity, and floral traits

Nectar content, plant height, visibility, and flower morphology all influence butterfly nectaring behavior. Tuberous, vividly colored blooms like those of *Clerodendrum* spp., *Ixora coccinea*, and *Asystasia gangetica* are preferred by species like *Papilio polytes* and *Delias eucharis* [21, 36].

While nectar specialists depend on a small number of species, many butterflies exhibit adaptability in their foraging habits, switching to common exotics when resources are scarce. For example, *Danaus genutia* has broad tolerance in nectar usage, visiting more than 20 kinds of flowering plants in disturbed environments [32, 42].

Seasons also influence floral tastes. Persistent bloomers like *Lantana camara* and *Bougainvillea spectabilis* dominate nectar foraging in the pre-monsoon, while wildflowers like *Tridax procumbens*, *Vernonia cinerea*, and *Cleome viscosa* are heavily relied upon during the monsoon months. Furthermore, research from urban West Bengal demonstrates that species such as *Eurema hecabe* and *Junonia orithya* modify their nectar consumption according to availability, underscoring the plasticity of nectar foraging tactics [41].

4.5 Generalist vs. specialist strategies in changing environments

Butterfly populations are changing as a result of the continuous changes to natural environments brought about by urbanization, agriculture, and climate change.

Because of their wide ecological amplitude, which includes tolerance to alien host and nectar plants as well as adaptability to disturbed habitats, generalists like *Danaus chrysippus*, *Catopsilia pomona*, and *Pieris rapae* exhibit resilience [33, 38].

Conversely, specialist butterflies show signs of decline. They are extremely susceptible because of their reliance on particular plant species and microhabitat conditions. Studies conducted throughout India and Southeast Asia have demonstrated population decreases for *Tanaecia lepidea*, *Zemeros flegyas*, and *Ariadne merione* in response to host plant displacement or fragmentation [34, 37].

So, it is essential to preserve ecological niches and plant diversity at the landscape level. Even in partially disturbed settings, butterfly diversity can be maintained through the conservation of host–nectar networks, which include native shrubs, herbs, and understory plants.

5. Synthesis, ecological implications, and conservation relevance

The link between plants and Lepidoptera provides a unique perspective on coevolution, ecological complexity, and biodiversity risks. Butterflies display extremely dynamic relationships with their hosts and nectar plants throughout their developmental phases. These relationships are influenced by anthropogenic influences, species characteristics, habitat structure, and environmental variability. These observations are summarized in this part to emphasize the significance of such plant–insect dynamics for ecology and conservation.

5.1 Ecological significance of host and nectar associations

In terrestrial environments, Lepidoptera serve two purposes: as pollinators as adults and as herbivores as larvae. Many species' reproductive success and geographic range are determined by host plant specialization, which is frequently the result of coevolutionary processes. For instance, *Zemeros flegyas* depends on *Maesa indica*, whereas *Ariadne merione* is only linked to *Trema orientalis* [20, 34]. These linkages show strong evolutionary ties with their host plants, which are tailored for oviposition cues, larval nutrition, and predator avoidance.

Generalist butterflies, on the other hand, such as *Catopsilia pyranthe* and *Eurema hecabe*, exhibit adaptable foraging and oviposition strategies by using a wider variety of hosts, including *Cassia tora*, *Senna occidentalis*, and *Trifolium repens* [7]. Nectar connections also mirror flower characteristics such as color cues, nectar sugar concentration, and corolla shape. While generalists like *Danaus chrysippus* and *Junonia orithya* exhibit nectar plasticity and regularly visit common weeds like *Tridax procumbens*, *Vernonia cinerea*, and *Lantana camara*, *Papilio demoleus* and *Delias eucharis*, for instance, favor tubular red or purple flowers like *Ixora coccinea* and *Clerodendrum infortunatum* [21, 32].

Particularly in fragmented or ecotonal environments, where butterflies frequently take the role of bees as important pollinators for small or open-flowered plants, these plant relationships support pollination, herbivory, and nutrient cycling [44].

5.2 Implications of habitat fragmentation and resource scarcity

Floral and larval supplies are less continuous and of lower quality when habitats are fragmented. Specialist species are disproportionately affected by this.

For example, the removal of native host trees such as *Trema orientalis* and *Ficus hispida* has been connected to the local extinction of *Ariadne merione* and *Tanaecia lepidea* in the Western Ghats and Bankura district [37, 41]. In a similar vein, host-dependent species like *Zemeros flegyas* have had their ability to reproduce restricted by the loss of understory herbs in tea garden environments.

By taking advantage of common nectar and larval supplies like *Cassia*, *Portulaca*, and *Calotropis* spp., generalists like *Junonia lemonias* and *Catopsilia pomona* demonstrate increased ecological adaptability and persist in semi-urban greenspaces and agro-ecological edges [33]. However, by displacing specialists and decreasing interaction diversity, their growth may lead to biotic homogenization, which would impair ecosystem resilience.

Furthermore, resource mismatch is made worse by phenological changes brought on by climate change. For example, pre-monsoon butterfly emergence and *Cleome viscosa* flowering asynchronously can decrease adult reproductive success and larval survival [24, 25].

5.3 Butterfly-plant interaction ecosystem and pollination services

Butterflies play significant ecological roles, particularly in disturbed areas where main pollinators are limited, despite being secondary pollinators in comparison to bees. They contribute to landscape-level gene flow in plants by their high mobility and cross-habitat foraging activity. For example, *Graphium agamemnon* and *Papilio polytes* pollinate *Lagerstroemia speciosa* and *Duranta erecta* even in urban green belts, while *Euploea core* is known to pollinate *Rauvolfia tetraphylla* and *Stachytarpheta jamaicensis* in both disturbed and forest edges [34, 45].

Foraging across field margins and bunds, *Pieris brassicae* and *Catopsilia pyranthe* promote nectar plants like *Ipomoea cairica* and *Tithonia diversifolia* in agroecosystems and aid in the reproductive success of crucifers and legumes. It is becoming more well acknowledged that these systems' nectar corridors are an essential part of long-term pollination services [13, 26].

5.4 Conservation implications and landscape management

Conservation of lepidopteran biodiversity requires landscape-level strategies that go beyond protected areas and encompass managed habitats like agroecosystems, urban greenspaces, and ecological corridors. Many butterflies have significant host specificity, which makes them especially sensitive to the loss of larval plants. The restoration of native host flora, such as *Trema orientalis* for *Ariadne merione* and *Maesa indica* for *Zemeros flegyas*, has been demonstrated to directly enhance the recolonization and stability of specialist populations.

In addition to larval requirements, butterflies rely on a variety of nectar sources, which might vary seasonally. Conservation plantings with staggered-blooming species like *Tridax procumbens*, *Vernonia cinerea*, and *Cleome viscosa* can provide constant adult feeding opportunities across seasonal gradients [10, 32].

Urban areas, which were once considered biodiversity deserts, are now being acknowledged for their conservation potential when planted with nectar-rich and native decorative species. Butterfly gardens with *Ixora coccinea*, *Catharanthus roseus*, and *Clerodendrum infortunatum* have been shown to support generalist species such as *Junonia orithya*, *Eurema hecabe*, and *Danaus chrysippus* in semi-urban areas of West Bengal and Uttar Pradesh [32, 41, 46]. Similarly, agroecological margins—such as field bunds and hedgerows—can operate as nectar corridors and microhabitats, increasing both pollinator variety and crop pollination spillover [13, 26].

Butterflies are also reliable indicators of ecological health. Long-term monitoring has revealed that species such as *Hypolimnas bolina* and *Tanaecia lepidea* are susceptible to microclimate variations, pesticide drift, and habitat simplification, making population trends useful markers for ecological assessment [36, 47]. Conservation frameworks should prioritize native floral diversity, temporal resource continuity, and habitat mosaics that support both larval and adult survival. Integrating these ideas into landscape management can improve butterfly resilience while also contributing to larger biodiversity goals in the face of climate change and anthropogenic disturbance.

6. Conclusion and future directions

Lepidoptera, which include both butterflies and moths, are among the most ecologically important insect taxa, serving as herbivores during their larval phases and pollinators as adults. Their interactions with plants involve both trophic levels and ecological processes, ranging from devouring host plant tissue to aiding cross-pollination. These interactions are determined by a mix of evolutionary fidelity and environmental availability. The evidence presented in this chapter demonstrates that these plant-insect connections are not random or uniform, but rather elaborately patterned by habitat type, seasonal change, and the landscape's ecological history. Butterfly-plant uniqueness evolved in response to evolutionary stresses such as chemical defense adaption, sensory cue recognition, and developmental synchrony. As previously discussed, many butterfly species are highly devoted to specific host plants (e.g., *Ariadne merione* to *Trema orientalis*, *Zemeros flegyas* to *Maesa indica*), whereas others are more adaptable in their nectar consumption, switching between a variety of blooming plants in response to temporal scarcity. These distinctions have significant consequences for conservation. Specialists are more vulnerable to local extinction if either their host or nectar sources dwindle, whereas generalists may survive but contribute to the ecological homogeneity of disturbed habitats. The findings highlight the importance of seeing butterfly conservation as a dynamic process that necessitates the temporal and spatial continuity of plant resources, rather than a single action or isolated intervention. This chapter offers a novel idea known as the Habitat-Specific Resource Continuity (HSRC) Framework. This model emphasizes the need of ensuring the continuous availability of both larval and adult floral supplies within specified habitat mosaics and throughout the seasonal cycle. It combines ecological uniqueness with landscape management to promote butterfly populations in both natural and artificial settings. The Habitat-Specific Resource Continuity (HSRC) Framework builds on the foundational idea that butterflies require not just "space" to live but a reliable and ecologically suitable sequence of plant resources over time. Unlike conservation models that focus solely on habitat area or species richness, HSRC highlights the need for spatiotemporal alignment between butterfly life cycle demands and floral resource availability. The model incorporates four key pillars:

Host Plant Fidelity – the degree to which larval host plants are present and undisturbed in a habitat.

Nectar Resource Succession – the availability of nectar plants across seasons, ensuring continuous foraging options.

Habitat Heterogeneity – the inclusion of microhabitats (e.g., forest edge, scrubland, grassland) that support varied ecological niches.

Landscape Connectivity – corridors or patches that allow movement between resource islands to ensure metapopulation stability.

This approach is not only philosophically sound, but it also applies to real-world conservation scenarios. For example, in a fragmented semi-urban zone, ensuring that nectar plants such as *Tridax procumbens*, *Cleome viscosa*, and *Catharanthus roseus* bloom in succession during different seasons can support generalist butterflies, while retaining or planting larval hosts such as *Calotropis gigantea* or *Passiflora suberosa* can aid in specialist survival. Similarly, on forest edges or rural landscapes, mixing natural nectar-rich plants with understory larval hosts can increase butterfly abundance and variety. Conventional conservation efforts frequently focus on the establishment or upkeep of protected areas. While necessary, they are insufficient, particularly in areas where human activity is predominant. The HSRC model instead encourages integrative design, in which habitat fragments, agroecosystems, school gardens, and urban parks are considered suitable butterfly habitats if plant diversity and resource continuity are preserved. Studies from West Bengal [41] and Karnataka [36] show that even modest urban green patches can support significant butterfly variety if floral and host plant layers are available all year. Urban planning and agroecological design should consciously include host and nectar plants. Native flowering plants with staggered blooming seasons, such as *Ixora coccinea*, *Vernonia cinerea*, and *Clerodendrum infortunatum*, can support adult butterfly populations, while larval hosts like *Cassia tora*, *Maesa indica*, and *Trema orientalis* promote breeding. In agricultural environments, keeping nectar corridors along bunds and uncultivated margins boosts ecosystem resilience and pollination services [26].

Importantly, the HSRC framework accounts for future ecological uncertainty. As climate change alters phenology, the timing of butterfly emergence and plant flowering may become more erratic. Restoration initiatives should prioritize variety in both plant species and bloom dates to provide ecological buffering against phenological mismatches. Implementing the HSRC paradigm necessitates collaborative efforts across ecological research, landscape management, and policymaking. First, long-term monitoring of butterfly populations in relation to plant phenology is required to validate resource continuity on a local scale. This includes citizen science projects, school biodiversity clubs, and mobile butterfly tracking apps that can provide massive datasets on butterfly-plant interactions.

Second, legislative incentives are required to incorporate butterfly conservation into municipal green space planning and agro-environmental initiatives. Incentives for farmers who keep flowering margins or schools who build butterfly gardens, for example, can improve habitat connectivity on a large scale. Third, more study is needed to map butterfly-plant networks throughout bioregions, particularly in little studied ecosystems such as mangroves, marshes, and high-altitude grasslands. Investigating how generalist and specialized butterflies interact with invasive plant species may help determine if these plants should be controlled or managed as temporary nectar sources.

Finally, education and community engagement must be the foundation of butterfly conservation. When locals understand that butterflies are not only decorative components but also bioindicators, pollinators, and ecological sentinels, they are more willing to help create butterfly-friendly surroundings. Butterfly conservation is more than just protecting insects; it is about sustaining the ecological connections that underpin biodiversity. This chapter has shown that the uniqueness of butterfly

DOI: http://dx.doi.org/10.5772/intechopen.1012048

host and nectar plant connections, as well as their sensitivity to seasonal and habitat variables, necessitate an ecologically nuanced and spatially aware conservation strategy.

The Habitat-Specific Resource Continuity (HSRC) Framework presented here provides an innovative and effective lens for guiding butterfly conservation across landscapes. It reconciles ecological theory with field realities, combining larval development, adult sustenance, seasonal cycles, and habitat fragmentation into a single integrated strategy. As the climate warms, landscapes fracture, and floral diversity declines, such a comprehensive and habitat-specific framework will be crucial not only for Lepidoptera but also for the resilience of the larger ecosystems they help sustain.

Author details

Avisek Patra
Department of Zoology, Bankura Christian College, Bankura, West Bengal

*Address all correspondence to: avisek@bankurachristiancollege.in

References

[1] Ohgushi T, Craig TP, Price PW, editors. Ecological Communities: Plant Mediation in Indirect Interaction Webs. Cambridge, UK: Cambridge University Press; 2007

[2] Thompson JN, Pellmyr O. Evolution of oviposition behavior and host preference in Lepidoptera. Annual Review of Entomology. 1991;**36**:65-89

[3] Kristensen NP, Scoble MJ, Karsholt O. Lepidoptera phylogeny and systematics: The state of inventorying moth and butterfly diversity. Zootaxa. 2007;**1668**:699-747

[4] Romeis J, Wäckers FL. Nectar- and pollen-feeding by adult herbivorous insects. In: Wäckers FL, van Rijn PCJ, Bruin J, editors. Plant-Provided Food for Carnivorous Insects. Cambridge, UK: Cambridge University Press; 2005. pp. 165-220

[5] Altermatt F, Pearse IS. Similarity and specialization of the larval vs. adult diet of European butterflies and moths. The American Naturalist. 2011;**178**(3):372-382

[6] van Swaay CAM et al. Global warming and excess nitrogen may induce butterfly decline by disrupting interactions. Ecology Letters. 2006;**9**(6):748-756

[7] Chew FS, Renwick JAA. Oviposition behavior in lepidoptera. Annual Review of Entomology. 1994;**39**:213-244. DOI: 10.1146/annurev.en.39.010194.001241

[8] Ehrlich PR, Raven PH. Butterflies and plants: A study in coevolution. Evolution. 1964;**18**(4):586-608

[9] Futuyma DJ, Agrawal AA. Macroevolution and the biological diversity of plants and herbivores. PNAS. 2009;**106**(43):18054-18061

[10] Tiple AD, Khurad AM, Dennis RLH. Butterfly larval host plant use in a tropical urban context. Journal of Insect Science. 2011;**11**(1):65. DOI: 10.1673/031.011.0101

[11] Dennis RLH et al. Habitat and resource use in butterflies: Towards a functional understanding of biodiversity. Biological Journal of the Linnean Society. 2011;**102**(1):1-21

[12] Alam MM, Rahman MM, Shuvo MH. Biology of Danaus chrysippus and feeding potentials in host and nectar plants. Journal of Biodiversity and Conservation. 2019;**3**(1):1-10

[13] Bahlai CA, Landis DA. Habitat type influences monarch butterfly (*Danaus plexippus*) oviposition. Environmental Entomology. 2019;**48**(3):675-682

[14] Severns PM, Breed GA. Behavioral consequences of exotic host plant use: Impacts of nectar and larval hosts on Fender's blue butterfly. Journal of Insect Conservation. 2014;**18**(3):381-393

[15] Wang X, Fu X, Shi M, Xue C, Yang J, Zhao Z, et al. Multiple interaction networks reveal that Lepidoptera larvae and adults prefer various host plants for diet and pollination. Integrative Zoology. 2024;**19**(4):763-776. DOI: 10.1111/1749-4877.12745

[16] Jia H et al. Pollen molecular identification from *Spodoptera exigua*. International Journal of Molecular Sciences. 2023;**24**(8):7588

[17] Ali JG, Agrawal AA. Specialist versus generalist insect herbivores and

plant defense. Trends in Plant Science. 2012;**17**(5):293-302

[18] Petschenka G, Agrawal AA. Milkweed butterfly resistance to plant toxins is linked to sequestration ability. Ecology Letters. 2015;**18**(3):339-346

[19] Futuyma DJ, Moreno G. The evolution of ecological specialization. Annual Review of Ecology and Systematics. 1988;**19**:207-233

[20] Forister ML, Dyer LA, Singer MS, Stireman JO, Lill JT. The global distribution of diet breadth in insect herbivores. PNAS. 2015;**112**(2):442-447

[21] Gilbert LE. Pollen feeding and reproductive biology of Heliconius butterflies. PNAS. 1972;**69**(6):1403-1407. DOI: 10.1073/pnas.69.6.1403

[22] Tudor M, Dennis RLH, Greatorex-Davies JN. The host-plant utilization of British butterflies. Journal of Insect Conservation. 2004;**8**(4):119-132

[23] Singer MC, Parmesan C. Phenological asynchrony and climate change. Philosophical Transactions of the Royal Society B. 2010;**365**(1555):3161-3176

[24] Boggs CL, Inouye DW. A single climate driver has direct and indirect effects on insect population dynamics. Ecology Letters. 2012;**15**(5):502-508

[25] Diamond SE, Frame AM, Martin RA, Buckley LB. Species' traits predict phenological responses to climate change in butterflies. Ecology. 2011;**92**(5):1005-1012

[26] Tscharntke T, Klein AM, Kruess A, Steffan-Dewenter I, Thies C. Landscape perspectives on agricultural intensification and biodiversity–ecosystem service management. Ecology Letters. 2005;**8**(8):857-874

[27] Bergman KO et al. Butterflies and habitat change in the Swedish countryside. Journal of Insect Conservation. 2004;**8**(4):133-143

[28] Van Dyck H et al. Declines in butterflies in Europe: A review of possible causes. Biological Conservation. 2007;**138**(1-2):1-20

[29] Öckinger E, Smith HG. Semi-natural grasslands as population sources for pollinating insects. Journal of Applied Ecology. 2007;**44**(1):50-59

[30] Biesmeijer JC et al. Parallel declines in pollinators and insect-pollinated plants in Britain and the Netherlands. Science. 2006;**313**(5785):351-354

[31] Van Swaay CAM, Nowicki P, Settele J, Van Strien AJ. Butterfly monitoring in Europe: Methods, applications and perspectives. Biodiversity and Conservation. 2010;**19**:3455-3469

[32] Singh A, Pandey R. Butterfly diversity in urban ecosystems of Central India. Urban Ecosystems. 2021;**24**:1237-1249

[33] Valtonen A, Saarinen K, Jantunen J. Rural–urban gradient and temporal variation in butterfly communities. Biological Conservation. 2014;**177**:141-150

[34] Kunte K. Butterflies of Peninsular India. Hyderabad, India: Universities Press; 2000

[35] Thomas JA. Monitoring change in the abundance and distribution of insects using butterflies and other indicator groups. Philosophical Transactions of the Royal Society B. 2005;**360**(1454):339-357

[36] Kunte K. Mimetic butterflies support Wallace's model of sexual dimorphism.

Proceedings of the Royal Society B. 2008;**275**(1644):1617-1624

[37] Ramesh T, Hussain KJ, Selvanayagam M. Patterns of diversity and habitat use of butterflies in a semi-arid region of southern India. Journal of Insect Conservation. 2014;**18**(5):847-864

[38] Haddad NM, Brudvig LA, Clobert J, Davies KF, Gonzalez A, Holt RD, et al. Habitat fragmentation and its lasting impact on Earth's ecosystems. Science Advances. 2015;**1**(2):e1500052

[39] Tiple AD, Khurad AM. Butterfly diversity in and around Nagpur city, Central India. Journal of Threatened Taxa. 2009;**1**(2):119-122

[40] Hanski I. Metapopulation Ecology. Oxford, UK: Oxford University Press; 1999

[41] Patra A, Patra A, Mandal B, Ghosh A. Diversity and abundance of butterflies in urban areas of Bankura district, Bankura, West Bengal, India. The Journal of Basic and Applied Zoology. 2025;**86**:10. DOI: 10.1186/s41936-025-00426-9

[42] Jain N, Verma R, Kamboj R. Urbanization and butterfly diversity: Patterns from Indian semi-arid landscapes. Urban Ecosystems. 2023;**26**:211-225

[43] Tiple AD, Khurad AM, Dennis RLH. Butterfly diversity and seasonal distribution in city parks of Nagpur, Central India. Journal of Environmental Biology. 2006;**27**(1):197-202

[44] Hegland SJ et al. How does climate warming affect plant–pollinator interactions? Ecology Letters. 2009;**12**(2):184-195

[45] Ariyanti NS, Fauzi AM, Iskandar J. Butterfly–nectar plant interactions in forest-agricultural transition zones of West Java. Biodiversitas. 2020;**21**(10):4648-4657

[46] Bahlai CA, Landis DA, Gratton C. Quantifying ecosystem services and disservices in agricultural landscapes: Pollination and pest control. Ecological Applications. 2019;**29**(6):e01968

[47] Knaden M, Hansson BS, Steck K. Butterfly responses to climate-driven changes in plant odor profiles. Nature Communications. 2022;**13**:5873

Chapter 4

The Role of Microstructures in Eclosion and Wing Expansion of Butterflies

Zhang Jinwen, Ling Xiaofei, Wang Weiwei, Lu Qin, Ding Weifeng, Han Shichang and Chen Hang

Abstract

Eclosion is the critical process of the butterfly transforming from a pupa into an adult. During this process, the morphology of the lepidopteran wings undergoes drastic changes, as the butterfly rapidly transitions from a wingless pupa to an adult with large-scale wings, resulting in a several-fold or even dozen-fold expansion in wing surface area. Although the lepidopteran wings undergo drastic morphological changes within a short period, this process does not cause damage to the butterfly's body. The reason lies in the pupal developmental stage, where the wing membrane, veins, tracheae, and membranous tissues gradually differentiate into numerous micron-scale foldable units. They are double-layered, foldable structures and capable of providing a several-dozen-fold expansion in surface area—far exceeding the actual requirement for wing expansion—thus supplying sufficient surface area reserves for wing flattening. During eclosion, the foldable units are progressively unfolded under wing expansion forces, with the stored area being utilized as the foldable units are flattened sequentially. At the macroscopic level, this results in rapid expansion of the surface area. The unfolding process of lepidopteran wings is influenced by the structure and arrangement of wing veins, exhibiting a change trend where length increases first, followed by width. Therefore, the morphological changes of lepidopteran wings are the result of cumulative deformations in foldable units. In summary, these foldable units serve not only as the constituent units of the wings but also as functional units, regulating wing deformation and the unfolding process.

Keywords: foldable units, eclosion, wing expansion, morphological changes, microstructures

1. Introduction

Butterflies are holometabolous insects, undergoing four distinct life stages—egg, larva, pupa, and adult—each stage has remarkable differences in morphology and behavior [1, 2]. The transformation from pupa to adult, known as eclosion, typically lasts 3–10 minutes [1–3]. Throughout the pupal stage, butterflies remain in a nearly motionless state, neither eating nor drinking, until eclosion is complete. Eclosion represents one of the most critical turning points in a butterfly's life. Through this process, butterflies develop lepidopteran wings—far larger than their own body size—and develop the ability to fly. This transformation provides essential mobility for survival and reproduction.

During the eclosion process, butterflies undergo profound transformations in morphology, structure, functionality, and behavior. The wings alone exhibit significant macroscopic changes, including: a several-fold increase in surface area; structural hardening from soft to rigid; scales from overlapping to spreading out; substantial reduction of hemolymph in wing veins; tracheal system expansion. These transformations occur within the brief 3- to 10-minute eclosion period, representing an irreversible, qualitative change in wing morphology (i.e., irreversible changes in surface area within an extremely short timeframe). For the organism, this represents a drastic mutational process. Without appropriate buffering mechanisms, butterflies would struggle to endure such radical physiological changes.

The sudden morphological transformation of lepidopteran wings represents a unique case in the developmental process of biological organs. Organ development in organisms typically requires an extended period, with rapidly formed organs often exhibiting weaker functionality, shorter lifespans, or insufficient stability. However, after their rapid formation, lepidopteran wings can withstand millions or more flapping cycles [4], while playing crucial roles in processes such as flight, signal transmission, sexual selection, warning displays, mimicry, and camouflage [5].

During the larval period, many structures of the lepidopteran wing are formed, and lepidopteran wing develop as imaginal discs [6–9]. Various genes control the change in the shape of lepidopteran wings. These genes activate corresponding signal proteins [10]. A layer of irregularly arranged transparent skin cells is observed after pupation. Old cells undergo apoptosis, and new cells are formed after cell division. Moreover, after about 32 hours, the cells are arranged normally [11]. These cellular changes cause physical torsion in the epithelial tissue, leading to deformation of the lepidopteran wing surface, which results in cell proliferation, apoptosis, growth, or morphological changes of lepidopteran wing [11, 12]. During the development stage, the single-layer membrane of lepidopteran wing turns over and finally folds to form a double-layer membrane, forming the back and abdomen of lepidopteran wing, respectively [13]. At this time, wings achieve a fixed shape [14, 15]. The cells near the middle of lepidopteran wing continue to elongate vertically and connect with the left and right cells to form cell clusters in the subsequent development process [11]. After 2–3 days of pupation, the lepidopteran wing cells of the pupa undergo programmed cell death [16, 17]. Then, the epidermal cells gradually degrade and heal into a wing membrane, which is covered with various sensory organs and body wall derivatives [18]. The body wall containing the trachea gradually develops into wing veins filled with various motor nerves, plasma, and hemolymph [19]. The cells among lepidopteran wing die, to form a transparent double-layer stratum corneum after eclosion. The wing vein cells survive, while the development of the epidermal cells stops, and hence the morphology, structure, function, pattern, etc. of lepidopteran wing complete their development.

Through in-depth research on lepidopteran wings, scientists have elucidated aspects such as genetic regulation, cellular development, wing shape, and pattern formation. However, studies on the physical structure and functional mechanisms during the eclosion process remain relatively limited. This chapter investigates the lepidopteran wing expansion process, analyzing morphological changes at the microstructural level to establish direct structure-function relationships. By doing so, it aims to uncover the fundamental mechanisms underlying rapid wing morphogenesis, thereby providing models and scientific data to advance research and applications of lepidopteran wing structures.

2. Materials and methods

2.1 The origin of butterflies

The experiment used *Kallima inachus* as the research subject, rearing 5th instar larvae until the butterflies emerged, and collecting wings from different days for scanning electron microscopy. The butterfly larvae were sourced from the Artificial Butterfly Breeding Garden at the Yuanjiang Experimental Station of the Highland Forestry Research Institute, Chinese Academy of Forestry (102°00′46″E, 23°36′11″N). The breeding garden employed outdoor rearing and indoor hatching methods for cultivation. When the *K. inachus* larvae is the 5th instar, they were transferred to an artificial climate chamber for rearing and fed on *Baphicacanthus cusia* (Nees) Bremek. The light period was set from 09:00 to 19:30, during which the temperature was maintained at 30°C and humidity at 70%. For the remaining time, the lighting was turned off, with the temperature set at 25°C and humidity at 70%.

2.2 Measuring morphological parameters of lepidopteran wings

The butterfly wings were carefully excised at their bases and air-dried. Subsequently, the wing specimens were mounted on an Ultra-Depth-of-Field 3D Microscope (Keyence VHX-1000) for imaging. When necessary, multiple overlapping images were acquired and automatically stitched using the built-in image stitching function to reconstruct complete wing images, particularly for larger specimens. The acquired images were then imported into CAXA 2007 software (Beijing Digital Heaven Co., Ltd.), where wing outlines were traced using key reference points. Morphometric parameters including wing length, width, and surface area were quantitatively analyzed using the software's measurement tools.

2.3 Sampling time

The time when the larvae just pupated was recorded as the starting point of the experiment (0 h). Every 24 hours, three samples were collected from the same batch of butterfly pupae until the butterflies fully emerged. The remaining pupae, which were not sampled, were used for eclosion and wing-spreading observations. The time when the pupae began to split was recorded as the eclosion starting point, and the timing and characteristics of the scales reaching their distinctive morphological features were documented.

2.4 Calculation of wing vein parameters

Measure the perimeter (L), area (S) of the wing veins, as well as the perimeter (l) and area (s) of the tracheae at each developmental stage. Then calculate their perimeter ratio (δ) and area ratio (η) using the following formulas:

$$\begin{cases} \delta_i = \sum \frac{L_{ij}}{l_{ij}} / n \\ \eta_i = \sum \frac{S_{ij}}{s_{ij}} / n \end{cases} \tag{1}$$

where i is the time period between samples, j is the sample number of the period, n is the number of samples in the period, L is the perimeter of a wing vein, S is the area of a wing vein, l is the perimeter of trachea, and s is the area of trachea. Relative area (Δs) is calculated as $\Delta s = S - s$.

3. Results and discussion

3.1 Macroscopic morphological changes of butterfly wings

When *K. inachus* initially pupates, the pupal already possesses lepidopteran wings on its abdomen. The structural development of lepidopteran wings primarily occurs during the pupal stage [20]. Throughout development, the wing morphology and surface area undergo dynamic alterations, with the transformation process divisible into two distinct phases: the pupal stage and the eclosion stage (**Figure 1**).

During the pupal stage, the morphology of the wing undergoes little change. Specifically, at 0 hours, the wing scales measured 12.89 mm in length, 9.53 mm in width, and 72.123 mm^2 in area. By 120 hours, these dimensions change to 12.68 mm in length, 9.22 mm in width, and 84.919 mm^2 in area. Over the 6-day pupal development period (0–120 h), the length, width, and area of the wings changed by factors of 0.98, 0.97, and 1.18, respectively.

During the eclosion period, the morphology of the lepidopteran wings undergoes drastic changes. At 120 h, the wing scales measured 12.68 mm in length, 9.22 mm in width, and 84.919 mm^2 in area. By 120 h 15 min, these dimensions increased to 32.37 mm in length, 31.91 mm in width, and 789.948 mm^2 in area. Over the 15-minute eclosion phase (120–120 h 15 min), the wing expanded by factors of 2.56 in length, 3.46 in width, and 9.30 in area.

The macroscopic morphological changes of wing primarily occur during the eclosion period. Comparative analysis of wing contour changes reveals that despite several-fold increases in area, length and width, the pre-eclosion and post-eclosion contours remain remarkably similar (**Figure 1**). Thus, the surface area expansion of wing scales is primarily occur during the eclosion period, representing an isotropic and homogeneous scaling process.

3.2 Microscopic morphological changes of butterfly wings

The lepidopteran wing is an integrated system composed of multiple structural components. Morphological changes in the constituent tissue structures necessarily induce corresponding changes in the lepidopteran wing morphology. The lepidopteran wing primarily consists of three key elements: the wing membrane, wing veins, and wing scales. The morphological changes in these constituent parts

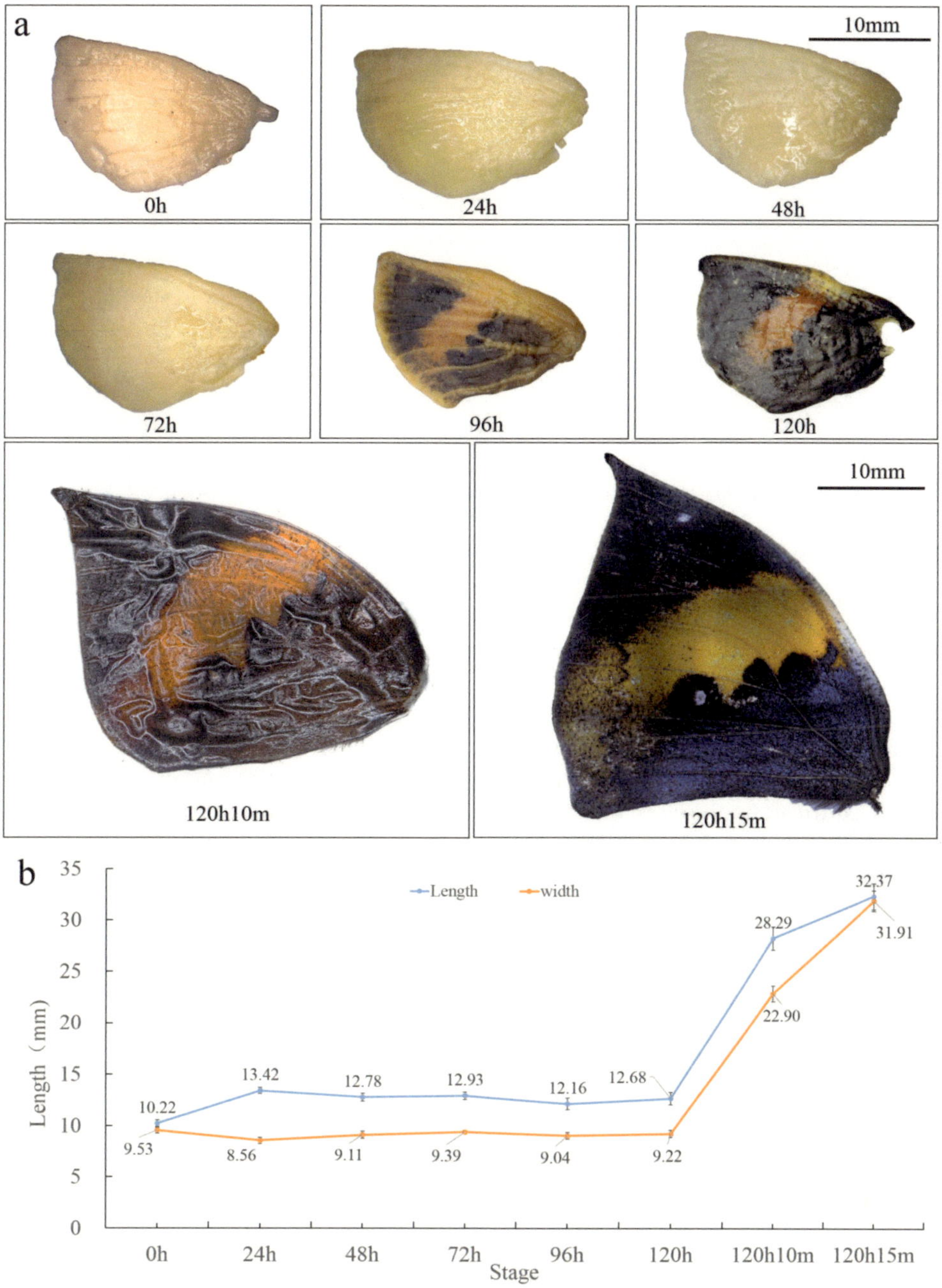

Figure 1.
Morphological changes of K. inachus *right forewings at different pupal stages. (a The morphological changes of the right forewing over time during pupation [21]. b Changes in the length and width of the right forewings.*

fundamentally determine the overall morphological changes of the lepidopteran wing. Accordingly, this study specifically investigates the morphological characteristics of these component structures (**Table 1**).

Stage	Area (mm^2)	Standard deviation (mm^2)
0 h	72.123	2.359
24 h	93.436	2.294
48 h	84.448	3.243
72 h	86.793	3.967
96 h	77.438	2.682
120 h	84.919	2.904
120 h 10 m	510.181	15.081
120 h 15 m	789.948	30.597

Table 1.
Variation in the surface area of wings.

3.2.1 Morphological changes of the wing membrane

The wing membrane is the largest surface area component of lepidopteran wing, accounting for approximately 91% of the total surface area (**Table 2**). As the primary structure governing wing morphological changes (**Table 2**). The wing membrane undergoes significant differentiation during pupal development, forming numerous cone-shaped microfoldable units. These interconnecting foldable units collectively constitute the wing membrane architecture, thereby microfoldable units as the primary structural elements of the wing membrane.

During the pupal stage, the morphology and quantity of the foldable units undergo continuous changes: at 0 h, the wings have not yet differentiated into forewings and hindwings. The wing membrane remains undivided into dorsal and ventral membranes, with no foldable units formed. At 24 h, the wings differentiate into forewings and hindwings, though the wing membrane remains undivided. Irregularly arranged foldable units begin to form on the membrane. At 48 h, the wing membrane differentiates into distinct dorsal and ventral membranes. At 72–120 h, cells in the dorsal and ventral membranes develop bundled structures, while the foldable units proliferate. At 120 h, the foldable units become aligned in orderly bands, with most units neatly arranged. The bundled structures stand vertically between the dorsal and ventral membranes.

Thus, over 6 days, the wing membrane accumulates a substantial reserve of microscopic foldable units—the primary structural components of the membrane. These foldable units exhibit a dual-layer folding configuration, serving as a reservoir of surface area for wing expansion during eclosion.

Name	Wing surface area (mm2)	Total length of wing veins Lc (mm)	Average width of wing veins Lb (mm)	wing veins area (mm2)	The ratio of wing veins
Forewing	789.95 ± 25.91	284.86 ± 4.98	0.22 ± 0.02	65.23 ± 7.11	8.26%
Hindwing	779.86 ± 34.58	269.46 ± 4.64	0.22 ± 0.02	61.98 ± 5.50	7.95%

Table 2.
Parameters of K. inachus *wings.*

3.2.2 Morphological changes of the wing veins

The wing veins constitute the second largest structural component of the lepidopteran wing, accounting for approximately 9% of the total wing area (**Table 2**). Wing veins comprise four distinct elements: the membranous matrices, the venation membranes, the tracheae, and the filamentous structures. Notably, three of these components – the membranous matrices, tracheae, and venation membranes – composed of the folding units (**Figure 2a, b, Table 3**).

Similar to the wing membrane, the folding units of the wing veins also develop progressively during the pupal stage. The developmental process is as follows: At 0 h, the vein membrane and membranoid substance have not yet formed; only tracheae are present within the wing structure. At 24 h, the vein membrane develops, but neither the vein membrane nor tracheae exhibit folding units at this stage. At 24–120 h, the number of folding units in the vein membrane and tracheae increases significantly. At

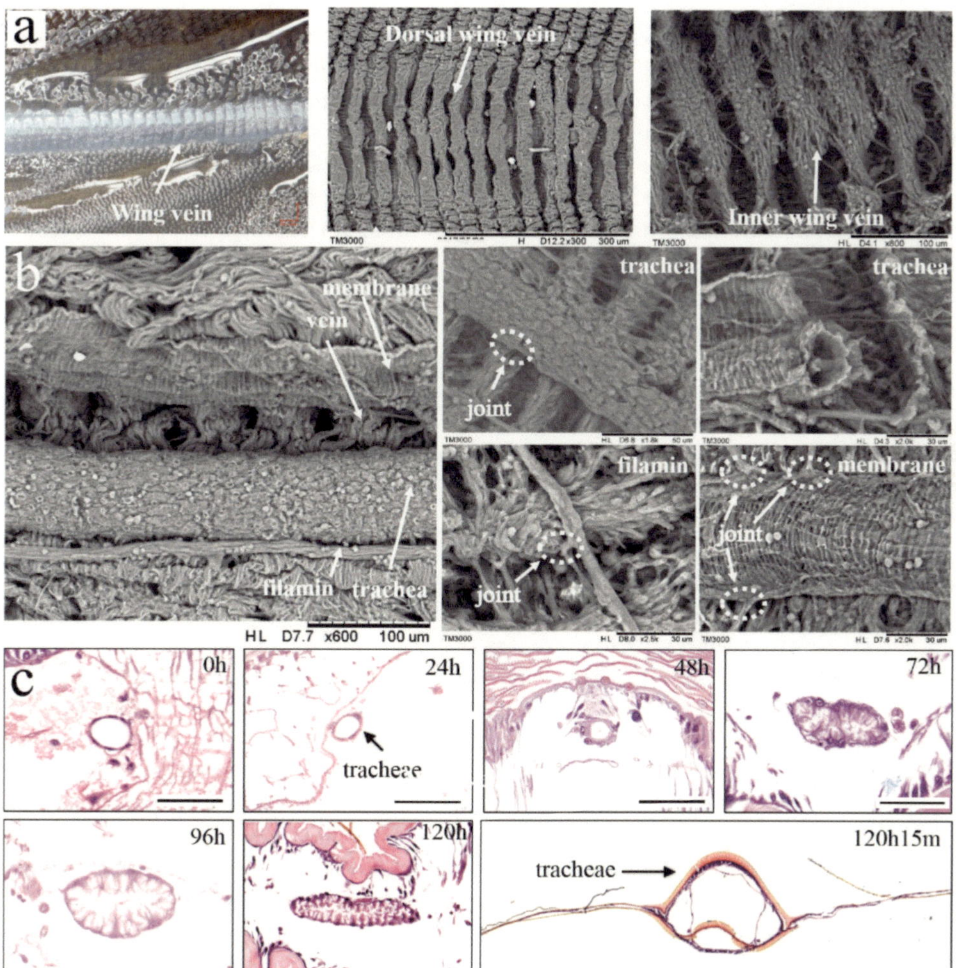

Figure 2.
Structure and development of wing veins. (a) External view of the folding units and wing vein. (b) The basic structure of wing veins. (c) The morphological change of trachea.

Stage	η(%)	δ(%)
0 h	5.277 ± 1.755	19.480 ± 8.32
24 h	12.238 ± 3.830	101.655 ± 14.520
48 h	11.915 ± 2.159	82.139 ± 14.502
72 h	3.483 ± 0.756	56.919 ± 9.812
96 h	3.116 ± 0.521	54.270 ± 8.879
120 h	1.845 ± 0.679	33.845 ± 8.17
120 h 15 min	1.152 ± 0.051	1.312 ± 0.070

Note: η is the ratio of the wing vein area to the trachea area. δ is the ratio of the wing vein perimeter to the trachea perimeter.

Table 3.
Parameters of K. inachus *wings.*

120 h 15 min, the folding units of the vein membrane and tracheae disappear, and dorsal and ventral membranes of these structures become tightly appressed (**Figure 2c**). Thus, the primary structures of the wing veins accumulate substantial quantities of microfolding units during the pupal stage. During development, the primary structures of the wing veins generate filamentous connections that integrate the veins and wing membrane into a unified system (**Figure 2b**). These filaments can ensure synchronized deformation among the primary structures of the wing.

3.2.3 Morphological changes of the bundled structures

The bundled structures are formed by the adhesion of filaments and are distributed between the dorsal and ventral wing membranes. These structures develop almost synchronously with the folding units, with their connection patterns varying according to location: those near the wing veins connect folding units within the same wing membrane, while those in other regions link folding units across different wing membranes.

The bundled structures gradually develop through the following stages: At 0 h, the bundled structures within the wing membrane do not formed. At 24 h, the bundled structures begin to form within the wing membrane. At 48–120 h, the number of bundled structures progressively increases. At 120 h 3 min, the bundled structures disperse, with filaments distributing between the dorsal and ventral wing membranes. At 120 h 15 min, the bundled structures completely disappear and form a defined interlayer within the wing membrane (**Figure 3a**).

The bundled structures show distinct functions at different developmental stages: during the pupal stage they primarily coordinate the quantity of folding units between dorsal and ventral wing membranes, during eclosion they mainly transmit wing-expansion forces while constraining the movement range of folding units, and in the adult stage they principally adhere the dorsal and ventral wing membranes together, making them essential structures for maintaining proportional area changes between both wing membranes.

The bundled structures serve distinct functions at different developmental stages: during the pupal phase they predominantly synchronize the proliferation of folding units between the dorsal and ventral wing membranes, throughout the eclosion period they mainly transmit wing-expansion forces while constraining the movement

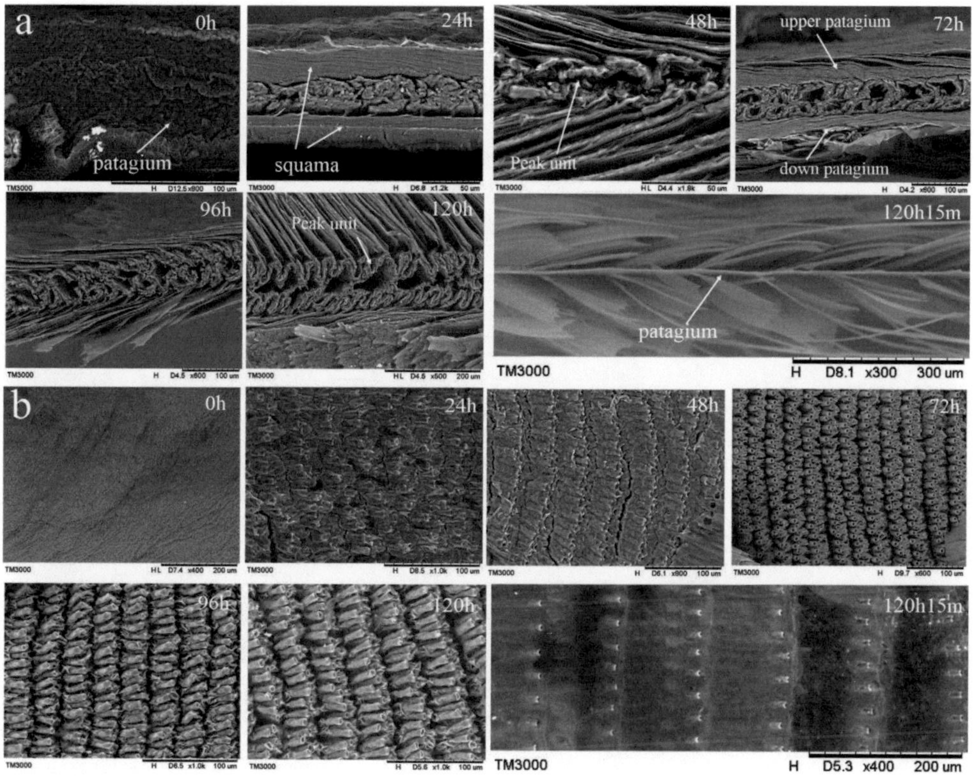

Figure 3.
Morphology and development of foldable units. (a) Cross-section mages of foldable units at different times. b Longitudinal section images of foldable units.

range of folding units, and in the adult stage they principally adhere the dorsal and ventral wing membranes together, making them essential structures for maintaining proportional area changes between dorsal and ventral wing membranes.

3.3 The expansion ways of butterfly wing

Wing expansion is a gradual developmental process characterized by four sequential phases: small and flat wings, small and wrinkled wings, elongated, slender wings, and large fully expanded wings. This deformation process follows a distinct bi-phasic pattern characterized by initial longitudinal elongation followed by subsequent lateral expansion (**Figure 4a, d**).

The constraint imposed by the pupal casing is responsible for the formation of this developmental pattern. During the pupal stage, the wings form numerous folding units (**Figures 2 and 3**). At eclosion, tracheal expansion generates the force for wing expansion, which is transmitted through bundled structures to the folding units. The sequential deformation of folding units leads to morphological changes in the wings (**Figure 4a–f**). Following eclosion, the hemolymph within the wings is progressively expelled, accompanied by progressive cellular apoptosis, ultimately resulting in the formation of a fully acellular wing.

During the process of emerging from the pupal case, the wings undergo longitudinal deformation under the combined effects of body movement and pupal case

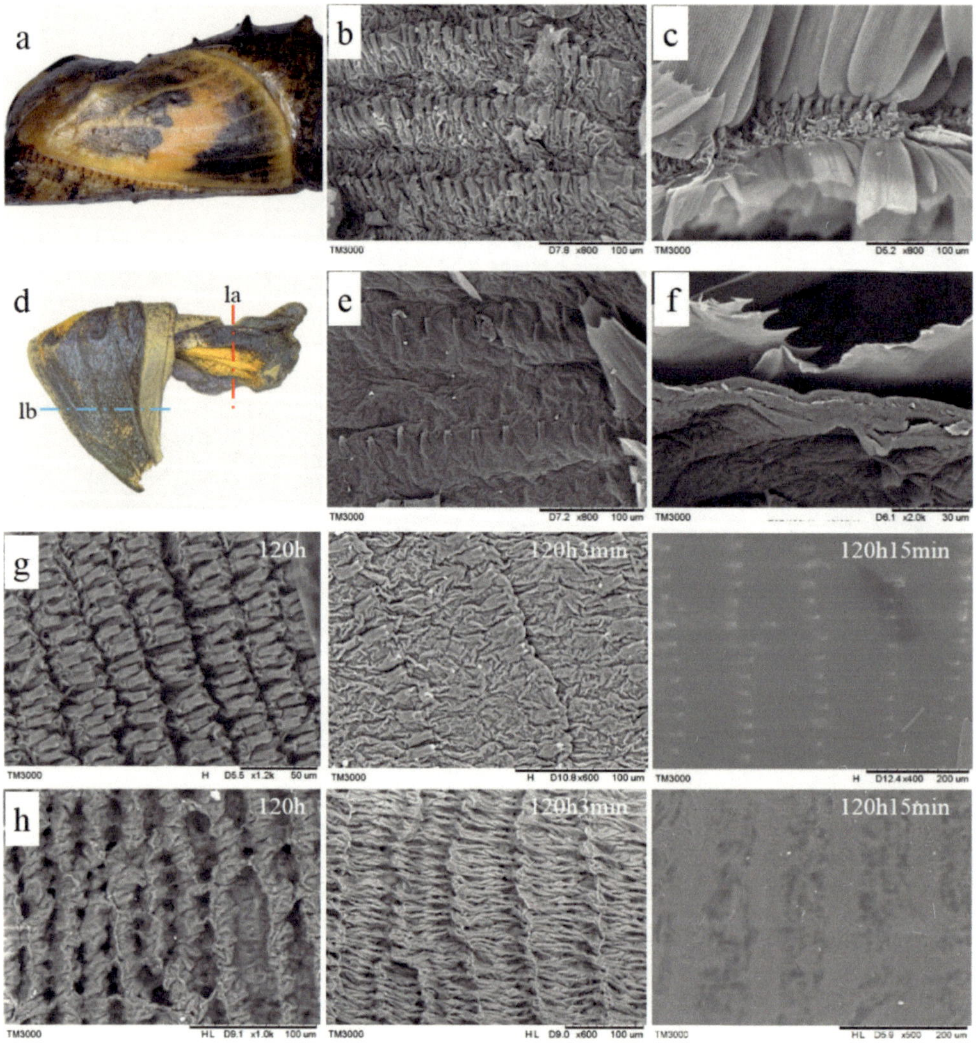

Figure 4.
The expansion sequence of the lepidopteran wing. (a) The wing in puparium. (b) The folding units at location la. c The cross-section of the lepidopteran wing at location la [21]. (d) The wing in emergence. la and lb are incisions [21]. e The folding units at location lb. f The cross-section of lepidopteran wing at location lb. g Morphological changes in the folding units of dorsal wing [21]. The butterfly pupa begins eclosion at 120 h and completes the process at 120 h 15 min. Hours: Morphological changes in the folding units of the inner wing.

constraint, followed by transverse deformation upon release from the pupal case constraint (**Figure 4d**). During pupal case shedding, the wing base is initially liberated from pupal case, triggering the unfolding of its folding units, while the wing apex is finally released with the subsequent unfolding of its folding units (**Figure 4b,e**). Therefore, the deformation of the lepidopteran wing follows a sequential base-to-apex progression pattern.

The deformation of the lepidopteran wing structure is a gradual process. During eclosion, a substantial amount of gas is generated within the tracheae, causing them to expand (**Figure 2b**). The expansion of the tracheae forces the wing veins to inflate, leading to morphological changes in the veins. The deformation of the veins is transmitted to the wing membrane *via* bundles, causing the wing membrane to undergo morphological changes as well. Throughout wing deformation, the folding units of the tracheae, venation membrane, and wing membrane are sequentially unfolded and bonded together by the bundles (**Figure 4g, h**). Since the deformation of the folding units is irreversible, the unfolding process of the wing is also irreversible.

4. Discussion

During eclosion, the morphology of the lepidopteran wings undergoes drastic changes, with its surface area rapidly expanding to 9.30 times the original size. This drastic changes results from the abundant reserves of microfoldable units accumulated during the pupal stage and the rapid unfolding of these structures during eclosion. The surface area of these microfoldable units is directly translated into the macroscopic wing surface area, thereby resulting in rapid expansion of the wing surface area.

The folding units undergo morphological changes during both the pupal and eclosion stages, yet the expansion of the wing occurs exclusively during eclosion. This distinction arises from fundamentally different transformation mechanisms between the two developmental phases: During the pupal stage, changes primarily involve the morphology, quantity, and size of the folding units; whereas in the eclosion stage, the predominant change is the unfolding of these pre-formed folding units. Consequently, this explains why the wing surface area shows minimal change during the pupal stage but exhibits rapid expansion during eclosion.

Wing expansion during eclosion is a precisely controlled process. Although the folding units of the dorsal and ventral wing membranes are asymmetric in size and quantity, the two membranes achieve perfectly matching area and shape upon full expansion. This is because the unfolding range of these folding units is constrained by connective bundles—while they can expand in any direction, their relative displacement cannot exceed the length of the bundles. When passively unfolded, the folding units lack sufficient driving force to overcome these mechanical constraints. Thus, both the degree and spatial distribution of their deformation are pre-determined.

The morphological changes of lepidopteran wings constitute a gradual transformation process. During eclosion, substantial gas production within the tracheae induces tracheal expansion, which compresses the hemolymph in wing veins and forces expansion of the veins. The expanded veins transmit wing-expansion forces through connective bundles to the wing membrane, triggering its morphological alteration.

Throughout wing deformation, the folding units of the tracheae, venation membrane, and wing membrane undergo sequential unfolding; these expanded folding units transform into membranous tissues and are interconnected by bundles, ultimately forming either wing membranes or veins. The pupal case constrains wing deformation during eclosion, with unconstrained regions undergoing initial deformation. Under the combined effects of body movement and pupal case restraint, wings first exhibit longitudinal elongation before developing widthwise expansion after cast-off from the pupal case. Microscopically, wing deformation follows a tracheae-to-venation-to-membrane sequence; macroscopically, it progresses from wing base to apex; dimensionally, longitudinal expansion precedes lateral extension.

In summary, these foldable units serve not only as the constituent units of the wings but also as functional units, regulating wing deformation and the unfolding process. The rapid unfolding of foldable units serves as the fundamental mechanism enabling both wings rapid deformation and surface area several-fold expansion in lepidopteran wings. Biomimetic investigation of wing structure morphology and functionality can provide physical prototypes and foundational datasets for novel space-deployable structures.

Acknowledgements

This work was funded by the Xingdian Talent Support Program (2023-1348), the Academician Expert Workstation in Yunnan Province (202405AF140107), and the Science and Technology Department of Yunnan Province (Grant no. 202449CE340005).

Declarations

The authors declare no conflicts of interest.

Author details

Zhang Jinwen[1,2,3], Ling Xiaofei[1,2,3], Wang Weiwei[1,2,3], Lu Qin[1,2,3], Ding Weifeng[1,2,3], Han Shichang[4] and Chen Hang[1,2,3]*

1 Institute of Highland Forest Science, Chinese Academy of Forestry, Kunming, China

2 Yunnan Key Laboratory of Breeding and Utilization of Resource Insects, Kunming, China

3 Key Laboratory of Protection and Utilization of Insects of National Forestry and Grassland Administration, Kunming, China

4 Faculty of Mechanical and Electrical Engineering, Kunming University of Science and Technology, Kunming, China

*Address all correspondence to: 627721388@qq.com

References

[1] Mcmahon DP, Hayward A. Why grow up? A perspective on insect strategies to avoid metamorphosis. Ecological Entomology. 2016;**41**(5):505-515

[2] Truman JW. The evolution of insect metamorphosis. Current Biology. 2019;**29**(23):1252-1268

[3] Piszter G, Kertész K, Horváth ZE, et al. Reproducible phenotype alteration due to prolonged cooling of the pupae of *Polyommatus icarus* butterflies. PLoS One. 2019;**14**(11):1-24

[4] Combes SA. Materials, structure, and dynamics of insect wings as bioinspiration for MAVs. Encyclopedia of Aerospace Engineering. 2010;7:1-10

[5] Tsai CC, Childers RA, Nan Shi N, et al. Physical and behavioral adaptations to prevent overheating of the living wings of butterflies. Nature Communications. 2020;**11**(1):551

[6] Nijhout HF, McKenna KZ. Wing morphogenesis in Lepidoptera. Progress in Biophysics and Molecular Biology. 2018;**137**:88-94

[7] Marcus JM. The development and evolution of crossveins in insect wings. The Journal of Anatomy. 2001;**199**(1-2):211-216

[8] Fujiwara H, Ogai S. Ecdysteroid-induced programmed cell death and cell proliferation during pupal wing development of the silkworm, Bombyx mori. Development Genes & Evolution. 2001;**211**(3):118-123

[9] Shimmi O, Matsuda S, Hatakeyama M. Insights into the molecular mechanisms underlying diversified wing venation among insects. Proceedings of the Royal Society B: Biological Sciences. 2014;**281**(1789):20140264

[10] Nienhaus U, Aegerter-Wilmsen T, Aegerter CM. In-vivo imaging of the drosophila wing imaginal disc over time: Novel insights on growth and boundary formation. PloS One. 2012;**7**:103-110

[11] Iwasaki M, Ohno Y, Otaki JM. Butterfly eyespot organiser: in vivo imaging of the prospective focal cells in pupal wing tissues. Scientific Reports. 2017;7(1):40705

[12] Masaki I, Yoshikazu O, Otaki JM, et al. Real-time in vivo imaging of butterfly wing development: Revealing the cellular dynamics of the pupal wing tissue. PLoS One. 2014;**9**(2):233-239

[13] Macdonald WP, Martin A, Reed RD. Butterfly wings shaped by a molecular cookie cutter: Evolutionary radiation of lepidotic wingn wing shapes associated with a derived Cut/wingless wing margin boundary system. Evolution & Development. 2010;**12**(3):296-304

[14] Le RC, Debat V, Llaurens V. Adaptive evolution of butterfly wing shape: From morphology to behaviour. Biological Reviews. 2019;**94**(4):1261-1281

[15] Polic D, Fiedler K, Nell C, et al. Mobility of ringlet butterflies in high-elevation alpine grassland: Effects of habitat barriers, resources and age. Journal of Insect Conservation. 2014;**18**(6):1153-1161

[16] Ohno Y, Otaki JM. Live cell imaging of butterfly pupal and larval wings in vivo. PLoS One. 2015;**10**(6):e0128332

[17] Zhou QX, Li YN, Shen XJ, et al. The scaleless wings mutant in Bombyx mori is

associated with a lack of scale precursor cell differentiation followed by excessive apoptosis. Development Genes and Evolution. 2006;**216**(11):721-726

[18] Pass G. Beyond aerodynamics: The critical roles of the circulatory and tracheal systems in maintaining insect wing functionality. Arthropod Structure & Development. 2018;**47**(4):391-407

[19] Chintapalli RTV, Hillyer JF. Hemolymph circulation in insect flight appendages: Physiology of the wing heart and circulatory flow in the wings of the mosquito *Anopheles gambiae*. Journal of Experimental Biology. 2016;**219**(24):3945-3951

[20] Dinwiddie A, Null R, Pizzano M, et al. Dynamics of F-actin prefigure the structure of butterfly wing scales. Developmental Biology. 2014;**392**(2):404-418

[21] Zhang JW, Chen XM, Lu Q, Liu JG, Ling XF, Wang WW, et al. Foldable units and wing expansion of the oakleaf butterfly during eclosion. Journal of Bionic Engineering. 2022;**19**(3):724-736

Chapter 5

Monitoring of Lepidoptera: Challenges and Insights

Ali Rajabpour and Fatemeh Yarahmadi

Abstract

Lepidopteran pests pose a major threat to global food security by causing extensive crop damage, necessitating effective monitoring and management strategies through Integrated Pest Management. This chapter examines traditional sampling methods—including absolute (mark-release-recapture) and relative (surveys, active and passive traps) estimation techniques—alongside emerging technologies such as smart traps, remote sensing, UAVs, and AI-driven predictive modeling, which enhance early pest detection and precision-based interventions. The integration of traditional and advanced approaches, supported by bioelectric sensors and bioacoustic monitoring, offers sustainable solutions for real-time pest tracking. However, further research is needed to optimize automated systems, enhance sensor accuracy, and develop affordable tools for small-scale farmers. By combining scientific advancements with practical applications, this chapter highlights pathways toward more sustainable pest management, ensuring agricultural productivity and food security in the face of growing global demand.

Keywords: population estimates, IPM, traps, remote sensing, precision agriculture

1. Introduction

1.1 Importance of lepidopteran pests in food security

Despite the decline in natural resources and rapid population growth in recent decades—particularly in developing countries—ensuring food security remains one of humanity's greatest challenges. The uneven global distribution of agricultural production has significantly impacted the political, economic, and social conditions of many nations. Powerful agricultural producers not only influence the economic and political fate of food-dependent countries but also dictate the daily lives of their populations [1].

Pests, diseases, and weeds are among the most critical threats to crops and other plants essential to human welfare (e.g., ornamental plants, pastures, and forests). Preventing crop damage from pests and diseases alone could save nearly four billion people from starvation [1].

Controlling agricultural pests—including vertebrates and invertebrates—is crucial, as they cause significant crop damage. Insects are considered as the main plant pests in agricultural and non-agricultural ecosystems. Numerous species of Lepidoptera (moths and butterflies) are key agricultural pests, causing significant damage to crops and non-cultivated plants worldwide. Their larval stages are typically the most destructive, feeding on leaves, fruits, stems, roots, and other plant parts, leading to both quantitative and qualitative losses. However, labeling these insects solely as pests would be unjust. They play vital ecological roles, particularly in pollination and nutrient cycling within ecosystems. Only when certain species reach high population densities do they become problematic, requiring targeted control measures. Therefore, it is critically important to monitor both the population status and damage levels of these pests through a precise surveillance program [1].

1.2 Importance of monitoring program

The successful implementation of an Integrated Pest Management (IPM) program fundamentally depends on establishing proper pest sampling protocols (through monitoring programs) and implementing effective forecasting systems. The data collected through these processes enable pest managers to accurately estimate pest density and determine whether populations have reached established intervention thresholds, for example, Economic Injury level and Economic threshold. IPM emphasizes systematic pest population monitoring through appropriate sampling methodologies to estimate both pest density and developmental status, facilitating informed decisions regarding more intensive control measures, including chemical interventions, based on scientifically validated action thresholds [2, 3]. This approach strategically complements other preventative pest management strategies.

Various sampling techniques have been developed for pest monitoring and forecasting, with methodology selection primarily determined by the target pest's habitat characteristics, behavioral patterns, and biological parameters. Optimal sampling protocols must achieve an appropriate balance between scientific accuracy and field practicality, ensuring both operational efficiency and cost-effectiveness. Consequently, developing a robust monitoring and forecasting program requires comprehensive understanding of multiple factors including all elements influencing pest biology and habitat requirements, specific host plant interactions, and the relevant socioeconomic context of the agricultural production system [1, 4].

Field-based monitoring and forecasting programs represent essential components for preventing pest outbreaks across diverse ecosystems including cultivated lands, forest systems, and rangeland environments. The temporal component of monitoring proves particularly critical for preventing economic damage, where well-designed sampling programs incorporating clearly defined decision-making levels enable agricultural producers and pest management specialists sufficient lead time to implement optimally timed control measures [1, 5].

Accurate prediction of insect population dynamics requires consideration of six fundamental parameters: population density (expressed as number of insects per unit area), spatial distribution patterns within the environment, birth rate dynamics, mortality rate factors, population age structure characteristics, and comprehensive analysis of population trends (quantifying changes in population size over temporal scales). These parameters collectively provide the necessary foundation for reliable population forecasting and informed management decisions [1, 5].

2. Sampling and population estimation in lepidoptera (butterflies and moths)

Sampling is a scientific method for collecting data on insect populations. This information supports pest management decisions, conservation efforts, and ecological research. Accurate population estimates are particularly important for predicting outbreaks and implementing IPM for Lepidoptera pests.

There are two main approaches to estimate Lepidoptera populations:

A. Absolute estimation: This method determines the exact number of butterflies or moths per unit area. While time-consuming, it can be done through (i) Subsampling (Counting larvae or adults in small sample areas (e.g., 1 m^2) and scaling up to larger areas), and (ii) Mark-Release-Recapture (MMR) methods. MRR methods work best in contained environments and can have significant margins of error [1, 4].

B. Relative estimation: Relative estimation involves recording insect counts per sampling unit without calculating absolute population density. This method compares populations across locations or time periods rather than determining exact pest densities. For instance, when evaluating three plant cultivars to identify which hosts fewer pest insects, sweep netting provides relative population estimates across treatment replicates, yielding comparative rather than absolute data [1, 4].

Sampling skill and consistency prove critical for reliable results. Standardized equipment and methods must be employed consistently throughout the study. The approach generates comparative data rather than absolute numbers, making it particularly valuable for IPM programs where relative differences matter more than exact counts [1, 4].

Common Relative Estimation Techniques

1. Field surveys (Transect walks): Researchers follow predetermined patterns through fields at consistent intervals, typically using X-shaped, zigzag, or U-shaped paths. A standard protocol involves walking an X-pattern across a field while taking uniform steps (e.g., every 5 paces), then counting insects within a defined quadrat frame or recording insects on plants within the sampling area (**Figure 1**) [1, 4].

2. Standardized sweep netting: This technique utilizes nets with specific dimensions—typically featuring 20 cm diameter hoops and 100-120 cm handles. Successful implementation requires maintaining strict consistency in several aspects: the sweeping technique must remain uniform across all sampling events, sampling effort should be equivalent across all plots, and timing should account for known insect activity patterns [1, 4].

Several factors significantly influence sweep netting results and must be carefully considered. Insect behavior and habitat preferences affect catch rates, as do plant height and density. The time of day, current weather conditions, and insect size and mobility all contribute to variations in sampling effectiveness. Researchers must document and account for these variables when interpreting results [1, 4].

Figure 1.
Wooden sampling quadrat frame for relative population estimation of Syringopais temperatella *(Lepidoptera: Gelechiidae) larvae in wheat fields [6].*

A. Use of sampling equipment: This section describes various sampling tools including different types of traps and suction devices. Traps are generally classified into two main categories: random (passive) traps and attractive (active) traps. Random traps, also known as passive traps, function by capturing insects that happen to come into contact with them through natural movement patterns. These traps do not employ any mechanisms to actively attract or direct insects toward them [1, 4]. Some passive traps include:

Water traps: These traps typically consist of shallow containers such as pans filled halfway with water. To reduce surface tension that might prevent insects from drowning (allowing their escape), a small amount of detergent is often added to the water. In some cases, these traps are combined with attractants like sex pheromones, light, or colored surfaces, making them non-random in function. However, their efficacy for relative population estimation of target insects must be validated through separate studies before deployment, as they may not be suitable for all species [1, 4].

Pitfall traps: Designed for ground-dwelling insects, these traps are particularly effective for estimating relative populations of species like the turnip moth (*Agrotis segetum* Schiffermüller & Denis, Lepidoptera: Noctuidae) (**Figure 2**). The larvae of this pest typically hide in shelters during the day and migrate toward plants at night, making pitfall traps a reliable tool for larval population assessment [7, 8].

Window or flight traps: Positioned along insect flight paths, these traps intercept flying insects. Upon collision with specialized panels, insects are redirected into a collection chamber, which may contain toxicants (**Figure 3**).

Attractive or active traps incorporate specific mechanisms designed to lure target insect species for sampling purposes. These traps exploit insects' natural tropic responses to various stimuli, or alternatively, their avoidance behaviors toward certain

DOI: http://dx.doi.org/10.5772/intechopen.1010716

Figure 2.
The figure illustrates the simple structure of a pitfall trap.

Figure 3.
Structure of a simple window trap for capturing flying moths.

environmental cues. The operational principle relies on the targeted species' innate reactions to particular attractants or repellents [1, 4].

Among the most commonly employed active traps are pheromone traps and light traps. Pheromone traps utilize volatile chemical compounds secreted by insects' specialized glands. These pheromones trigger behavioral responses in other individuals of the same species. Of these, sex pheromones represent the most important category for relative population estimation of lepidopteran in monitoring programs. Sex pheromones are volatile compounds typically secreted by one sex (often females) and detected through chemoreceptors located primarily on the antennae of the opposite sex. Remarkably, these chemical signals can elicit attraction responses over considerable distances, with only a few airborne molecules sufficient to trigger behavioral reactions in receiving individuals [1, 9]. Commercial manufacturers have developed synthetic versions of these compounds after identifying their chemical structures. The synthesized pheromones are encapsulated in small dispensers for field use. The standard implementation involves placing these pheromone capsules within specialized trap designs, such as delta and funnel traps. The attracted adult insects become captured on adhesive panels positioned beneath the dispenser (**Figure 4**). These pheromone capsules have finite operational lifespans, as the active compounds gradually evaporate over time, necessitating periodic replacement. Environmental variables, particularly temperature and humidity, significantly influence capsule longevity. Moreover, the adhesive panels may lose effectiveness due to dust accumulation and also require regular replacement.

Pheromone traps serve crucial roles in monitoring adult emergence patterns and flight activity peaks of lepidopteran insects. By correlating peak flight periods with established biological data for specific pests, these traps enable optimal timing of chemical control measures. For instance, research on the pomegranate fruit borer (*Spectrobates ceratonia* Zeller, Lepidoptera: Pyralidae) in Ilam Province demonstrated that larval hatch occurs approximately 2–3 weeks after peak moth captures in pheromone traps. This finding allows growers to time insecticide applications to target vulnerable first-instar larvae before they penetrate the fruit through the calyx region [10]. Several environmental and technical factors influence the effectiveness of pheromone traps in capturing Lepidoptera, including trap height, wind direction and speed, rain, ambient temperature, relative humidity, inter-trap spacing, trapping duration and location, and trappers' experience [11]. The effectiveness of synthetic pheromones in attracting insects can vary greatly based on the chemical profile of the pheromone extract and its geographical origin [12]. These variables must be carefully considered and standardized prior to trap deployment to ensure reliable monitoring results.

Light traps represent another important category of active traps, exploiting the positive phototaxis exhibited by many insect species. These traps prove particularly

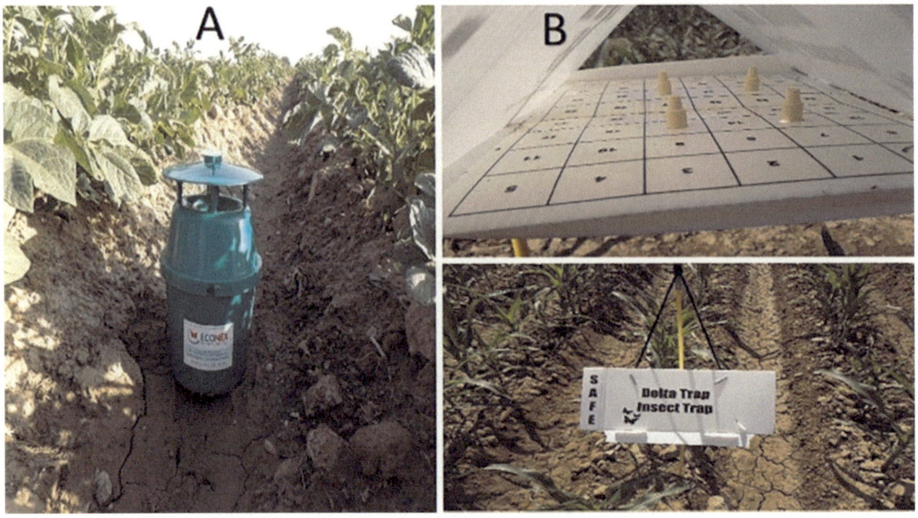

Figure 4.
The funnel (A) and delta traps containing pheromone capsule for capturing Phthorimaea operculella *Zeller and* Spodoptera exigua *Hubner in potato and corn fields, respectively.*

valuable for monitoring moths and enabling relative population estimates in pest surveillance programs. The effective light spectrum varies among species, requiring careful selection based on the target insect's visual sensitivity. It is demonstrated that shorter-wavelength artificial light disproportionately attracts larger moth species with bigger eyes and bodies, potentially disrupting ecosystems by affecting pollination and food webs. The findings reveal a size bias in light traps and suggest using longer-wavelength lighting to reduce ecological impacts [13].

Light trap surveys of nocturnal moths require careful standardization to account for key environmental variables. Based on continuous trapping over 225 nights in eastern Germany (yielding 372 species from 49,472 individuals), three critical factors emerged: temperature, humidity, and light source characteristics significantly affected nightly capture rates. For short-term surveys (<10 nights), sampling during the warmest summer nights proved most effective, while extended monitoring benefited from targeting warm nights distributed throughout the active season (March–October) [14].

The distance between light traps significantly influences moth sampling efficiency. Research shows that low-power light traps (such as those equipped with two 15 W UV tubes) have a very limited attraction radius, typically less than 10 meters. These findings confirm that captured moths primarily originate from the immediate local habitat rather than being attracted from distant areas. For accurate monitoring of habitat quality and fragmentation patterns, traps should be spaced at appropriate intervals (exceeding their attraction radius) to avoid oversampling the same local population. This limited attraction range makes moths particularly reliable ecological indicators for assessing habitat health [15].

Manufacturers have developed various light trap designs optimized for different agricultural settings, including distinct models for field crops versus orchard systems. Combining multiple attractants can significantly improve trapping efficiency. For instance, the Ferolite trap integrates light, color, and pheromones to maximize capture rates of target pests. This multi-stimulus approach has proven effective for monitoring populations of key lepidopteran pests, such as *Tuta absoluta* Meireck [16], *A. segetum* [7], *Syringopais temperatella* Led. (Lepidoptera: Gelechiidae) [6], and *Spodoptera exigua* Hubner (Lepidoptera: Noctuidae) (**Figure 5**) [17].

Figure 5.
Ferolite trap for trapping Spodoptera exigua *Hubner in sugar beet field.*

The strategic deployment of these active traps provides critical data for IPM programs, especially for monitoring Lepidoptera species with strong flight activity. Selection of appropriate trap types must consider the target species' specific behavioral ecology as well as local environmental conditions that may influence trap efficiency. When properly implemented, these monitoring tools contribute significantly to timely and effective pest management decisions while reducing unnecessary pesticide applications.

Smart traps for tracking moth pests: For generations, farmers have struggled with moths and caterpillars that damage their crops. Recently, smart trap technologies have revolutionized pest control by providing real-time data about pest populations. These advanced monitoring tools help farmers protect their fields more effectively while reducing unnecessary pesticide use. Modern smart traps come in several different designs, each suited for specific types of moths. Pheromone traps work by attracting male moths using synthetic scents, then automatically counting them with built-in sensors or cameras [18]. Light traps take a different approach, using UV lights to lure night-flying moths while artificial intelligence helps identify the species [19]. Another type, suction traps, operates quietly by pulling insects into a container for later examination. For storage facilities, some traps even use sound detection to recognize moths by their distinctive wingbeats [20].

What makes these traps truly "smart" is their connectivity. They can send pest counts and alerts directly to a farmer's smartphone or computer, providing instant updates. Some models go a step further by including weather sensors that help predict when pest populations might surge.

The shift to smart traps offers farmers several important benefits. First and foremost, they save significant time by eliminating the need for daily manual trap inspections. Instead, the system automatically sends alerts only when pest levels reach concerning thresholds. This efficiency translates to cost savings, as farmers can precisely target their pesticide applications rather than spraying entire fields unnecessarily. Beyond immediate practical benefits, these traps provide valuable long-term data. By tracking pest populations over time, farmers can identify patterns and predict outbreaks before they occur. Perhaps most importantly, this targeted approach to pest control means fewer chemicals in the environment, benefiting both ecosystems and farm workers.

While promising, smart trap technology is not without its challenges. The initial investment can be substantial, particularly for small-scale farmers. Rural operations may face additional hurdles with power supply and internet connectivity needed to run these systems effectively. Like any technology, smart traps require regular maintenance to function properly, and environmental factors like heavy dust or rain can occasionally interfere with sensor accuracy.

Sampling should be conducted in a manner that provides sufficient accuracy for decision-making while minimizing time and cost. To achieve this, two key factors must be considered: the sampling technique (the method used to count insects within a sampling unit) and the sampling plan (the strategy used to collect samples and estimate populations). An effective sampling plan should define the sampling unit, determine the optimal sampling area, estimate the required sample size for adequate precision, establish the spatial arrangement of sampling units, and identify the best timing for sampling [1, 4].

The sampling unit refers to a defined portion of the insect's habitat where population counts are conducted.

Samples collected from inappropriate sites (such as leaves or roots) would fail to accurately represent the pest's presence and population density. These units must be distinct and non-overlapping. The form of a sampling unit can vary significantly—examples include a specific number of leaves, a measured length of a branch, a one-square-meter wooden quadrat, or even 20 consecutive sweeps of an insect net across a crop row. The key requirement is that the sampling unit must allow for easy and accurate counting. For instance, a standardized quadrat can serve as an effective sampling unit. Smaller sampling units generally facilitate quicker and more cost-effective insect counts while reducing errors. For example, counting aphids on a single leaf is far simpler and less prone to error than counting aphids on an entire tree. However, excessively small sampling units may introduce estimation errors, particularly at low population densities, due to insufficient representation of the habitat [1, 4].

The identification of optimal sampling areas is crucial for minimizing sampling errors and preventing data misinterpretation. Each species' distinct behavioral and biological characteristics determine their primary activity locations, which in turn define the most appropriate sampling areas. The sampling unit must correspond precisely to the pest's habitat. For example, when monitoring codling moth (*Cydia pomonella* L.), the larval stage inhabits fruits, while the pupal stage resides in bark crevices. Therefore, sampling units must be aligned with these specific locations. The location of larval activity may vary significantly depending on both the developmental stage and seasonal timing. A notable example is the peach twig borer (*Anarsia lineatella* L.), whose early-instar larvae exhibit shoot-feeding behavior prior to fruit formation. This behavioral shift necessitates corresponding adjustments in sampling methodology to ensure accurate population monitoring.

Research conducted in corn fields provides a clear example of how sampling areas vary between species. For the corn stem borer (*Sesamia cretica* Led.), the optimal sampling area was found to be the mid-plant section, while for the beet armyworm (*S. exigua*), sampling was most effective in the lower plant section [21]. These differences highlight the importance of species-specific sampling strategies. Host plant characteristics also play a significant role in determining optimal sampling areas, as they directly influence herbivore behavior and biology. This was demonstrated in a study of the potato tuber moth (*Phthorimaea operculella* Zeller; Lepidoptera: Gelechiidae) conducted on two different potato cultivars, Santea and Arinda. The research revealed distinct optimal sampling areas for each cultivar. Furthermore, the study showed that optimal sampling areas varied depending on the target developmental stage. For egg monitoring specifically, the upper leaves showed the highest correlation with total plant egg counts, while no significant correlation was found in the lower leaves [22]. These findings underscore the necessity of considering multiple factors when establishing sampling protocols, including species-specific behaviors, host plant characteristics, and developmental stages. Such comprehensive consideration ensures accurate pest population monitoring and reliable data collection for effective pest management strategies.

Determining the appropriate sample size is essential for conducting an accurate and cost-effective sampling program. While increasing the sample size improves precision, excessive sampling can be expensive and time-consuming. Thus, the optimal sample number should be determined based on population characteristics [1, 4]. Sampling precision refers to the variability around the estimated population mean or the coefficient of variation—distinct from accuracy, which measures how close the estimate is to the true population value [23]. The desired precision level is a key factor in determining sample size, with higher precision requiring more samples.

Various equations help calculate the required sample size, using parameters like mean, variance, standard deviation, and standard error, often obtained through preliminary sampling. Below are some key formulas:

1. Fixed Standard Error Approach as Eq. (1).

$$n = \left(\frac{SD}{E \times \mu}\right)^2 \tag{1}$$

where, SD = standard deviation, μ = population mean, and E = acceptable standard error (e.g., 5% for IPM studies)

2. Poisson & Negative Binomial Distributions as Eqs. (2) and (3):

$$N = \frac{1}{E^2 \times \mu} \tag{2}$$

$$n = \frac{\left(\frac{1}{\mu} + \frac{1}{k}\right)}{E^2} \tag{3}$$

where, k = repetition count in the distribution

3. Confidence Limit-Based Calculation as Eq. (4):

$$n = \left(\frac{t \times SD}{D \times \mu}\right)^2 \tag{4}$$

t = t-value (1.96 for 95% confidence), and D = half-width of confidence interval

4. Taylor's Power Law (Spatial Dispersion) as Eq. (5):

$$n = t^2 \times a\overline{x}^{b-2} \times D^{-2} \tag{5}$$

where, a, b = regression parameters of Taylor's Power Law

5. Infestation Probability (Binary Data) as equation as Eq. (6):

$$n = \frac{t^2 \times p \times q}{D^2} \tag{6}$$

P = infestation probability, q = 1 - p

6. Coefficient of Variation (CV) method as Eq. (9):

$$SE = \frac{SD}{\sqrt{n}} \tag{7}$$

$$CV = \frac{SE}{\mu} \tag{8}$$

$$n = \frac{S^2}{\mu^2 \times CV^2} \tag{9}$$

Stratified Sampling for Different Habitats: When pest distribution varies across habitat subdivisions (e.g., citrus leaf caterpillar on stems vs. leaves), stratified sampling ensures accuracy. The required sample size per unit (e.g., trees in an orchard) is presented in Eq. (10):

$$n_t = \frac{\left(\frac{S_S^2}{n_s}\right) + S_p^2}{(\mu + SE)^2} \tag{10}$$

S_S^2 = within-unit variance, and S_p^2= between-unit variance.

This approach minimizes systematic errors from biased pest distribution [4].

Sequential sampling optimizes pest monitoring by dynamically adjusting sample sizes based on observed data, reducing time and costs compared to fixed-size sampling.

There are two main approaches to sequential sampling models:

1. Classification-based (e.g., Iwao's method, Wald's SPRT)—Determines if pest density exceeds a critical threshold (e.g., Economic Injury Level, EIL).

2. Fixed-precision estimation (e.g., Green's and Kuno's models)—Estimates pest density with a predefined accuracy (e.g., precision levels D = 0.25 for IPM, 0.1 for research).

3. Classification sequential sampling

Iwao's Method
Decision Boundaries Eqs. (11) and (12):
Upper:

$$U_n = n \times \left(cd + z_{\alpha/2}\sqrt{S^2/n}\right. \tag{11}$$

Lower:

$$U_n = n \times \left(cd + z_{\alpha/2}\sqrt{S^2/n}\right. \tag{12}$$

cd stands for critical density (also called the critical decision density or action threshold).

Sampling stops if cumulative counts (S_n) cross a boundary or reach maximum samples.

Wald's SPRT (Sequential Probability Ratio Test).

Tests hypotheses:

(H_0): Pest density *cd*.

(H_1): Pest density > cd.

Decision Lines: Parallel straight lines based on likelihood ratios.

Terminate sampling S_n crosses a boundary.

Error Rates: α (Type I), β (Type II) typically set at 0.05–0.10.

4. Fixed-precision sequential sampling:

This method estimates pest density with a specified precision (D).
Green's Model (Taylor's Power Law) Eqs. (13) and (14)

Sample Size:

$$n = \frac{1}{D^2 \times a\mu^{(b\;\,2)}} \tag{13}$$

Stop Line:

$$T_n \geq \left(\frac{an^{1-b}}{D^2}\right)^{\frac{1}{(2-b)}} \tag{14}$$

Kuno's Model (Iwao's Patchiness Regression) as Eqs. (15) and (16)

Sample Size:

$$n = \frac{1}{D^2 \times \left[\frac{(\alpha+1)}{(\mu+(\beta-1)}\right]} \tag{15}$$

Stop Line:

$$T_n \geq \frac{(\alpha+1)}{\left[\frac{D^2-(\beta-1)}{n}\right]} \tag{16}$$

In both models, T_n is expressed as the cumulative pest number, and the stop sampling lines can be illustrated by plotting T_n against n. The estimated stop lines imply that sampling from the population can be stopped when T_n reaches or exceeds the line. The b or β and a or α are the slope or intercept in Taylor's power law and Iwao's patchiness regressions, respectively [4].

Studies indicate that factors such as host plant variety [22] and weed conditions [21] in the field influence the required sample size calculated by sequential sampling models for some lepidopteran larvae. It is essential to recognize that host plants directly influence the biological traits of herbivores, including their spatial distribution—a fundamental parameter in sequential sampling models [24].

3. The role of precision agriculture in monitoring lepidopteran pests

Precision agriculture plays a transformative role in monitoring lepidopteran pests by enabling targeted and data-driven pest management strategies. Advanced technologies within this framework provide unprecedented accuracy in detecting and controlling these pests, fundamentally changing traditional monitoring approaches. Remote sensing (RS) and Geographic Information Systems (GIS) allow for early detection of larval infestations by analyzing multispectral imagery that reveals subtle changes in crop health. There are many studies that use RS and GIS to detect early infestations of lepidopteran pests on various host plants. They use different vegetation indices for this detecting. For instance, in oat jungles of west Iran, the Normalized Difference Vegetation Index (NDVI) was derived for the test area using Landsat 5, 7, and 8 imageries. Since Landsat 8 OLI's red and near-infrared bands differ from the others, a calibration model was developed. This model proved effective, with a high correlation coefficient (0.928) and low root mean square error (0.05). The NDVI showed a notable decline over the study period, likely due to *Tortrix viridana* L. (Lepidoptera: Tortricidae) infestation, given the area's protected status and minor temperature variations [25]. Greene et al. [26] examined spatial associations between three key lepidopteran pests—velvetbean caterpillar (*Anticarsia gemmatalis* Hübner), soybean looper (*Chrysodeixis includens* Walker), and green cloverworm (*Hypena scabra* Fabricius)—and soybean health indicators in South Carolina fields. Results demonstrated that NDVI exhibited stronger correlations with pest distributions than plant height or defoliation measurements, suggesting its potential as an early detection tool. Spatial analysis revealed significant pest-plant associations in 14% of cases, with most occurring during the critical first four weeks of sampling. These findings highlight NDVI's value for identifying emerging pest hotspots and support the development of targeted monitoring strategies early in the growing season when management interventions may be most effective. Another research developed a monitoring and forecasting framework for *Lymantria dispar* L. defoliation in Western Siberia using Sentinel-2-derived NDVI. During the 2017 outbreak, NDVI at two defoliated sites dropped significantly (-0.101 and -0.087), with spatial analysis revealing heavy defoliation (up to 10% forest area) at low elevations (916–1466 m) near roads and slopes. From 2017 to 2021, defoliation declined as forests recovered. A Dynamic Mode Decomposition model captured normal vegetation dynamics, with negative NDVI deviations signaling defoliation [27].

RS-based models can predict the timing of pest outbreaks using vegetation indices such as NDVI. By analyzing vegetation cover changes, this approach enables early identification of at-risk areas and supports optimized pest management. For example, Soukhovolsky et al. [28] developed novel distance indicators to proactively identify stands losing resistance to Siberian silk moth (*Dendrolimus sibiricus* Tschetv.) by analyzing their response to seasonal weather fluctuations rather than absolute values. Remarkably, these indicators revealed changes 2–3 years prior to actual outbreaks, enabling early risk assessment. The approach demonstrates that remote sensing can generate complex resistance indicators for global forests, transforming large-scale monitoring by predicting pest outbreaks before irreversible damage occurs. Scientists found that satellite data tracking snowmelt and spring vegetation growth can predict outbreaks of certain moths better than temperature data alone. For two spring-active moth species in Finland, these natural events signaled peak activity times more accurately. While the method did not work for other moths, it allowed researchers to create nationwide forecast maps. This approach could help farmers and foresters prepare for pest outbreaks by using satellite observations of seasonal changes as early warnings. The technique works best for pests whose life cycles match visible landscape changes [29].

Meanwhile, IoT-based sensors deployed across fields can detect adult moth activity through pheromone traps or acoustic monitoring, providing real-time population data. Machine learning (ML) algorithms enhance predictive capabilities by processing historical climate data and pest life cycle information to forecast outbreak risks, while decision support systems (DSS) generate customized management recommendations based on dynamic economic thresholds. A key advantage of precision agriculture lies in its ability to reduce pesticide use through localized interventions. Big Data analytics further optimize treatment timing by identifying the most vulnerable pest growth stages (e.g., first-instar larvae) for maximum control efficiency. However, challenges remain, particularly for species with irregular spatial distribution patterns (e.g., cotton bollworms), requiring specialized algorithms for accurate monitoring. Additionally, integrating field-collected data (e.g., pheromone or light trap counts) with automated smart systems remains critical for robust pest surveillance. To make pest monitoring easier and more precise, researchers developed a smart insect trap (E-trap) that uses a pheromone lure and a wireless camera to automatically count captured moths including *Spodoptera litura* Fabricius (Lepidoptera: Noctuidae) daily. Unlike traditional traps that require manual checks every five days, this high-tech version sends real-time images over 4G networks, letting farmers track pest numbers remotely without frequent field visits. When tested in Japanese soybean fields, the E-trap caught the same pest trends as old-school traps but revealed finer details—like daily population swings—which manual methods could miss. Even better, the system's AI-powered counts matched real insect numbers with over 95% accuracy, proving it's both reliable and labor-saving. This innovation could revolutionize pest control by giving farmers faster, data-driven insights while cutting down on tedious fieldwork [30]. Another Automated Moth Trap (AMT) that uses light lures and a camera to snap pictures of live insects at night was developed in Denmark. A smart AI system, called Moth Classification and Counting (MCC), then analyzes the images to identify species and count moths without harming them. In field tests over 48 nights, the trap captured 250,000+ images (averaging 5675 per night), and a custom deep learning model achieved 93% accuracy in species recognition. While tracking performance had room to improve (71–79% precision), the system proved to be a low-cost, hands-off way to monitor moths—giving scientists and farmers better data without the tedious fieldwork [31]. The electronic funnel trap (e-funnel) was developed for automated monitoring of pheromone-responsive Lepidoptera species. This system incorporated an optical counting mechanism and utilized a LoRa-based wireless network to transmit insect counts along with Global Positioning Systems (GPS) coordinates, timestamps, and temperature data to a cloud server. Field evaluations comparing the e-funnel network with conventional traps for monitoring *T. absoluta* populations demonstrated statistically equivalent counting accuracy between automated and manual methods, though the electronic traps captured approximately 84% of the insects collected by traditional plastic traps. The technology provided reliable real-time monitoring capabilities while eliminating the need for manual trap inspections, representing a significant advancement in automated pest surveillance [32].

RS technology has become an indispensable tool for monitoring plant health status; detecting nutritional needs, irrigation requirements, and soil conditions; and identifying biotic and abiotic stressors such as pests, diseases, and weeds. Through spaceborne, airborne, and drone-based geospatial data acquisition, RS enables accurate crop loss assessment and serves as a critical component in IPM decision-making systems. Recent advancements in AI and machine deep learning (MDL) have significantly enhanced the efficacy of RS data interpretation, facilitating more precise and timely management decisions [4].

As a rapid, large-scale observation tool, RS technology processes aerial and satellite imagery to deliver comprehensive data on pest habitats, damage severity, and population density dynamics. This capability is particularly valuable for area-wide pest interventions, where successful suppression relies heavily on spatially extensive data for IPM planning. AI and MDL play a pivotal role in analyzing the vast datasets generated by RS sensors, enabling cost-effective, rapid, and accurate predictions for informed decision-making. GIS, GPS, and RS function as complementary technologies, collectively enhancing the collection, integration, and analysis of spatial big data. These tools allow for the evaluation of pest population fluctuations across both

DOI: http://dx.doi.org/10.5772/intechopen.1010716

temporal and spatial dimensions, providing a more holistic understanding of infestation patterns. While traditional ground-based monitoring remains essential, RS and aerial data-gathering technologies offer spatially continuous insights into pest occurrence and dispersal over large regions—a capability particularly crucial for invasive or migratory species [4].

As an example, researchers developed an efficient drone-based system to detect destructive bagworm (Lepidoptera: Psychidae) infestations in oil palm plantations, addressing the limitations of traditional labor-intensive monitoring methods. Using a specialized DJI Inspire 2 drone flying at 70 meters altitude and equipped with a high-resolution Micasense Altum-PT multispectral camera, the team successfully identified four distinct infestation severity levels: healthy, low, mild, and severe. The study revealed that different combinations of vegetation indices performed best for specific detection tasks—the NDVI and Green Normalized Difference Vegetation Index (GNDVI) combination perfectly identified healthy and low-infestation areas, while NDVI with NDRE worked best for detecting moderate and severe cases. Among all indicators tested, NDVI emerged as the most consistently reliable across all infestation levels. The system's remarkable accuracy was further enhanced by the Weighted KNN algorithm, which achieved near-perfect detection rates exceeding 99.7%. This innovative aerial monitoring approach provides plantation managers with a rapid, large-scale solution for early pest detection, enabling timely and targeted control measures while significantly reducing labor requirements compared to conventional ground surveys [33].

A cost-effective unmanned aerial vehicles (UAVs) monitoring system using RGB imagery and machine learning (Random Forest and CNN) was developed to detect *Erannis jacobsoni* Djak infestations in Mongolian larch forests through vegetation indices and canopy texture analysis. Achieving over 85% accuracy comparable to multispectral methods, this approach enables rapid local pest detection while establishing a framework for scalable forest surveillance, providing managers with an efficient tool for early outbreak intervention and ecosystem preservation [34]. The detrimental impact of the bagworm *Metisa plana* Walker on Malaysian oil palm plantations was demonstrated through its capacity to cause 10–13% leaf defoliation and up to 40% yield losses, prompting the development of an efficient UAV-based monitoring system to overcome the limitations of time-consuming manual censuses; however, the inherent data imbalance in UAV imagery posed classification challenges for determining infestation severity levels, which were addressed through evaluation of three vegetation index combinations (NDVI-NDRE, NDVI-GNDVI, NDRE-GNDVI) coupled with various resampling techniques (ROS, SMOTE, RUS, 3-interval and 5-interval undersampling), revealing that 3-interval undersampling of imbalanced data achieved optimal performance with 86.84% classification accuracy and 100% F1-score, while Fine KNN consistently excelled in classifying all infestation levels using NDVI-NDRE features across datasets, with the counterintuitive finding that a 66.67% sample size reduction enhanced classification success even with unbalanced data, ultimately establishing an effective framework for large-scale pest monitoring in oil palm plantations through strategic data balancing approaches [35]. In Brazil, research aimed to detect leaf miner (*Leucoptera coffeella* Guérin-Méneville & Perrottet) infestations in newly planted coffee crops using vegetation indices (VI) derived from multispectral aerial images captured by a remotely piloted aircraft (RPA). The research applied the random forest (RF) algorithm for classification. The RF model in R software classified infested and healthy plants, revealing significant spectral differences. Vegetation indices also varied significantly: EXR was higher in

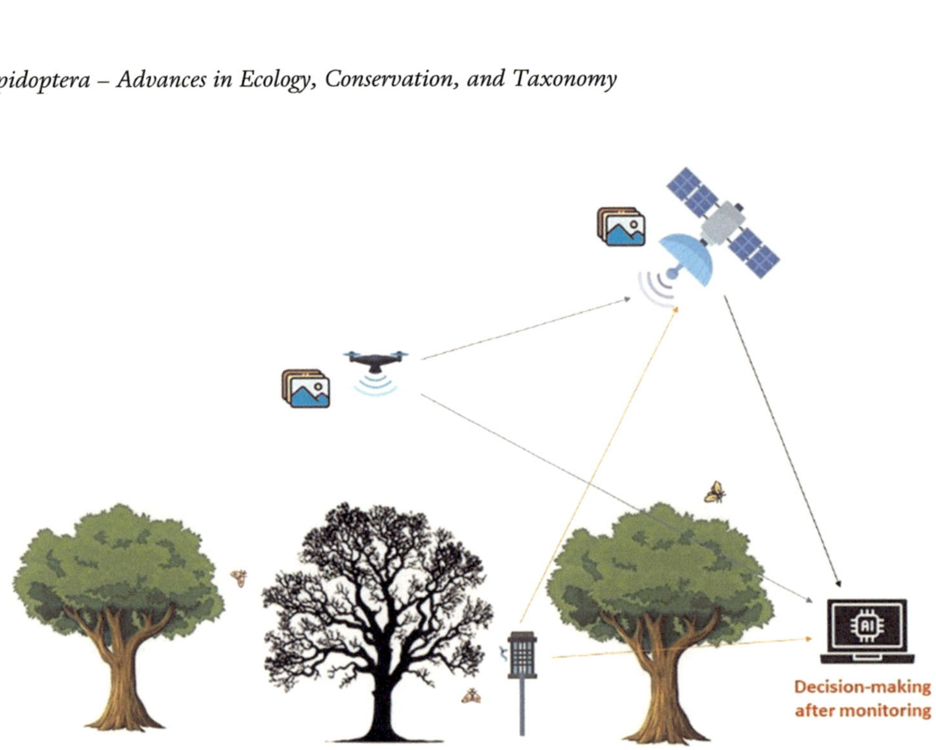

Figure 6.
The diagram illustrates the data flow between various monitoring components, including satellite and aerial imaging (via drone technology), Smart IoT-enabled traps, and AI-powered data analysis for decision-making. All collected data are transmitted to a central processing unit where artificial intelligence algorithms evaluate pest population dynamics and support management decisions.

infested plants, while GNDVI, GOSAVI, GRRI, MPRI, NDI, NDRE, NDVI, and SAVI were higher in healthy plants, indicating different light absorption patterns. Due to these spectral differences, the RF algorithm performed effectively in classification [36].

Ground monitoring alone is often time-consuming, costly, and impractical, especially in remote or difficult-to-access ecosystems such as forests, rangelands, and other natural habitats [37]. RS bridges this gap by delivering scalable, real-time data that support proactive and sustainable pest management strategies. By integrating RS with AI-driven analytics, modern IPM systems can achieve unprecedented precision in pest detection, monitoring, and control, ultimately optimizing agricultural productivity while minimizing environmental impact (**Figure 6**) [4].

4. Bioelectric sensors for lepidopteran insect monitoring

Bioelectric sensors are devices that detect and measure the electrical activity of living organisms, such as insects. These sensors are designed to capture the tiny electrical signals that are generated by the body's physiological processes, including sounds and chemical cues. The sensors can play a critical role in real-time monitoring of the lepidopteran insects in different ecosystems in the future. For instance, insects have a highly developed sense of smell that allows them to detect volatile chemicals released by plants or another insect, which they use to find food or mates. By combining an insect's sense of smell with electronic instruments, researchers have created a "bioelectronic nose" that can detect plant damage. This innovative technology uses a field-effect transistor connected to an insect antenna to create a stable bioelectronic interface. In a real-world test, the bioelectronic nose was used to detect a specific volatile chemical released by damaged plants in a glasshouse or another mate. The results were impressive, with the system able to detect the chemical at very low levels, even in the presence of 1000 undamaged plants. The detection was also extremely fast, taking only a few seconds to detect the chemical. This technology has the potential to be a powerful tool for monitoring and detecting plant damage and could have significant implications for agriculture and conservation [38].

Traditional methods of monitoring, such as hand netting and pan traps, are time-consuming and labor-intensive. To address this, researchers have been exploring the use of bioacoustic sensors to monitor flying insects. These sensors detect the sounds made by insects, such as buzzing, and can provide valuable insights into their populations. A study in Italy compared the effectiveness of bioacoustic sensors with traditional monitoring methods. The results showed that the sensors were able to detect changes in insect abundance and activity patterns, particularly in relation to temperature. While the sensors were effective in estimating overall insect abundance, further development is needed to accurately identify specific insect species. The use of bioacoustic sensors offers a promising solution for easing the burden of insect monitoring, which is essential for conservation efforts. However, it's important to note that these innovative technologies should not replace the expertise and data quality provided by professionals. Instead, they can be used to supplement traditional methods and provide unprecedented opportunities for monitoring and conservation. By combining traditional methods with new technologies, researchers and conservationists can work together to protect and preserve biodiversity [39]. For instance, it is demonstrated that some moths respond to ultrasonic waves at determined frequencies [40, 41]. Therefore, the findings can be used for developing bioacoustic sensors for detecting the insects.

5. Conclusion

Effective monitoring and population estimation are fundamental to implementing successful IPM strategies, ensuring timely interventions while minimizing unnecessary pesticide use. This chapter highlights the importance of robust sampling methodologies—including absolute and relative estimation techniques—to accurately assess lepidopteran pest densities. Traditional methods such as field surveys, sweep netting, and pheromone/light traps remain essential, but emerging technologies like smart traps, remote sensing, and AI-driven precision agriculture are revolutionizing pest surveillance. The integration of GIS, RS, and UAVs enables large-scale, real-time monitoring of pest infestations, improving early detection and targeted control measures. Moreover, advancements in bioelectric sensors and bioacoustic monitoring offer promising future tools for non-invasive, automated pest tracking. However, challenges such as high initial costs, technical maintenance, and species-specific variability in trap efficacy must be addressed to ensure widespread adoption.

Ultimately, a combination of traditional and innovative approaches—supported by data-driven decision-making—will enhance the sustainability and efficiency of lepidopteran pest management. By leveraging precision agriculture, machine learning, and IoT-based solutions, farmers and researchers can optimize pest control strategies, reduce ecological impacts, and safeguard agricultural productivity in an era of growing food demand. Future research should focus on refining automated monitoring systems, improving sensor accuracy, and developing cost-effective solutions for small-scale farmers to ensure equitable access to advanced pest management technologies.

Acknowledgements

We extend our gratitude to Agricultural Sciences and Natural Resources University of Khuzestan for providing the facilities necessary for writing this book chapter. The authors acknowledge the use of DeepSeek for language polishing of this chapter.

Conflict of interest

The authors declare that there is no conflict of interest.

Acronyms and abbreviations

IPM	integrated pest management
MMR	mark-release-recapture
EIL	economic injury level
CV	coefficient of variation
RS	remote sensing
GIS	geographic information systems
NDVI	normalized difference vegetation index
ML	machine learning
AI	artificial intelligence
GPS	global positioning systems
UAVs	unmanned aerial vehicles
KNN	the k-nearest neighbors' algorithm
GNDVI	green normalized difference vegetation index
GOSAVI	green optimized soil adjusted vegetation index
GRRI	green red ratio index
MPRI	modified photochemical reflectance index
NDI	normalized difference index
NDRE	normalized difference red edge index
SAVI	normalized difference red edge index

Author details

Ali Rajabpour* and Fatemeh Yarahmadi
Department of Plant Protection, Faculty of Agriculture, Agricultural Sciences and Natural Resources University of Khuzestan, Bavi, Iran

*Address all correspondence to: rajabpour@asnrukh.ac.ir; ali.rajabpour@mq.edu.au

References

[1] Rajabpour A. Monitoring and Forecasting in Plant Pest Management. Bavi: Agricultural Sciences and Natural Resources University of Khuzestan Press; 2022. 201 p

[2] Pedigo LP, Buntin GD, editors. Handbook of Sampling Methods for Arthropods in Agriculture. New York: CRC Press; 1993. 736 p

[3] Binns MR. Sequential sampling for classifying pest status. In: Pedigo LP, Buntin GD, editors. Handbook of Sampling Methods for Arthropods in Agriculture. Boca Raton: CRC Press; 1994. pp. 137-174

[4] Rajabpour A, Yarahmadi F. Decision System in Agricultural Pest Management. Singapore: Springer Nature; 2024. 368 p

[5] Leather SR, Watt AD. Sampling theory and practice. In: Leather SR, editor. Insect Sampling in Forest Environments. London: Blackwell publishing; 2006. pp. 1-15

[6] Rashidi D, Rajabpour A, Zandi-Sohani N. Fixed-precision sequential sampling plan of *Syringopais temperatella* (Lep., Gelechiidae) in wheat fields of Iran. Journal of Agricultural Science and Technology. 2020;**22**(6):1501-1510

[7] Heibatian AH, Yarahmadi F, Abadi LJ, A. Field efficacy of biorational insecticides, azadirachtin and Bt, on *Agrotis segetum* (Lepidoptera: Noctuidae) and its carabid predators in the sugar beet fields. Journal of. Crop Protection. 2018;7(4):365-373

[8] Yarahmadi F, Dinarvan N. Potential of resistance inducers for controlling *Agrotis segetum* Denis & Schiffermüller (Lepidoptera, Noctuidae) in sugar beet in Khuzestan. Iran Entomonlogical. 2022;**47**(4):469-472

[9] Dent D. Insect pest management. 2nd ed. Ascot: CABI Publishing; 2000. 410 p

[10] Yaghobi S, Rajabpour A, Zandi-Sohani N. Seasonal population dynamics, sampling distribution, and fixed-precision sequential sampling of *Spectrobates ceratoniae* (Lepidoptera: Pyralidae) in pomegranate orchards. Journal of Economic Entomology. 2020; **113**(3):1513-1518

[11] Pawlina I, Gilbert P. Factors affecting trap efficiency: A review. In: Proulx G, editor. Mammal Trapping. 1999. pp. 95-116

[12] Hong SY, Yi HJ, Yoon YN, Jang YW, Do Park K, Maharjan R. Evaluation of commercial pheromones on the population dynamics of *Spodoptera frugiperda* (JE smith) and *Mythimna loreyi* (Duponchel) (Lepidoptera: Noctuidae). The Korean Journal of Crop Science. 2022;**67**(4):285-295

[13] Van Langevelde F, Ettema JA, Donners M, WallisDeVries MF, Groenendijk D. Effect of spectral composition of artificial light on the attraction of moths. Biological Conservation. 2011;**144**(9):2274-2281

[14] Jonason D, Franzén M, Ranius T. Surveying moths using light traps: Effects of weather and time of year. PLoS One. 2014;**9**(3):e92453

[15] Truxa C, Fiedler K. Attraction to light- from how far do moths (Lepidoptera) return to weak artificial sources of light? European Journal of Entomology. 2012;**109**(1):77-84

[16] Nazarpour L, Yarahmadi F, Saber M, Rajabpour A. Short and long term effects

of some bio-insecticides on *Tuta absoluta* Meyrick (Lepidoptera: Gelechiidae) and its coexisting generalist predators in tomato fields. Journal of Crop Protection. 2016;**5**(3):331-342

[17] Darabian K, Yarahmadi F. Field efficacy of azadirachtin, chlorfenapyr, and *Bacillus thuringensis* against *Spodoptera exigua* (Lepidoptera: Noctuidae) on sugar beet crop. Journal of the Entomological Research Society. 2017;**19**(3):45-52

[18] Suto J. Codling moth monitoring with camera-equipped automated traps: A review. Agriculture. 2022;**12**(10):1721. DOI: 10.3390/agriculture12101721

[19] Bjerge K, Nielsen JB, Videbæk Sepstrup M, Helsing-Nielsen F, Høye TT. A light trap and computer vision system to detect and classify live moths (Lepidoptera) using tracking and deep learning. bioRxiv. 2020;**21**(2):343. DOI: 10.1101/2020.03.18.996447

[20] Aslam A, Ijaz MU, Aslam MT, Chattha MU, Khan I, Gulzar MZ, et al. Monitoring and detection of insect pests using smart trap technologies. In: Haq ZU, Ali I, editors. Revolutionizing Pest Management for Sustainable Agriculture (). New York: IGI Global; 2024. pp. 443-468

[21] Dinarvand N, Rajabpour A, Sohani NZ, Farkhari M. Effect of weedy culture on population densities, spatial distributions and sampling procedures of *Spodoptera exigua* and *Sesamia cretica* (Lep., Noctuidae) in corn fields. Bulletin of Entomological Research. 2020;**110**(1): 84-95

[22] Shahbi M, Rajabpour A. A fixed-precision sequential sampling plan for the potato tuberworm moth, *Phthorimaea operculella* Zeller (Lepidoptera: Gelechidae), on potato cultivars. Neotropical Entomology. 2017; **46**:388-395

[23] Henderson PA, Southwood TRE. Ecological Methods. 4th ed. John Wiley & Sons; 2016

[24] Mohammadi S, Seraj AA, Rajabpour A. Evaluation of six cucumber greenhouse cultivars for resistance to *Tetranychus turkestani* (Acari: Tetranychidae). Journal of Crop Protection. 2015;**4**(4):545-556

[25] Gooshbor L, Bavaghar MP, Amanollahi J, Ghobari H. Monitoring infestations of oak forests by *Tortrix viridana* (Lepidoptera: Tortricidae) using remote sensing. Plant Protection Science. 2016;**52**(4):270. DOI: 10.17221/185/2015-PPS

[26] Greene AD, Reay-Jones FP, Kirk KR, Peoples BK, Greene JK. Spatial associations of key lepidopteran pests with defoliation, NDVI, and plant height in soybean. Environmental Entomology. 2021;**50**(6):1378-1392

[27] Mori N, Kawatsu K, Noriyuki S, Kosilov A, Martemyanov V, Yamashita M, et al. Monitoring and prediction of the spongy moth (*Lymantria dispar*) outbreaks in mountain's landscape using a combination of Sentinel-2 images and nonlinear time series model. Forest Ecology and Management. 2024;**563**: 121975

[28] Soukhovolsky V, Kovalev A, Goroshko AA, Ivanova Y, Tarasova O. Monitoring and prediction of Siberian silk moth *Dendrolimus sibiricus* Tschetv. (Lepidoptera: Lasiocampidae) outbreaks using remote sensing techniques. Insects. 2023;**14**(12):955

[29] Pöyry J, Böttcher K, Fronzek S, Gobron N, Leinonen R, Metsämäki S, et al. Predictive power of remote sensing

versus temperature-derived variables in modelling phenology of herbivorous insects. Remote Sensing in Ecology and Conservation. 2018;**4**(2):113-126

[30] Kawakita S, Sato T. Towards automatic monitoring of insect pests using IoT camera-equipped pheromone traps: A case study for *Spodoptera litura* (Lepidoptera: Noctuidae). Applied Entomology and Zoology. 2023;**58**(3): 265-272

[31] Bjerge K, Nielsen JB, Sepstrup MV, Helsing-Nielsen F, Høye TT. An automated light trap to monitor moths (Lepidoptera) using computer vision-based tracking and deep learning. Sensors. 2021;**21**(2):343

[32] Rigakis II, Varikou KN, Nikolakakis AE, Skarakis ZD, Tatlas NA, Potamitis IG. The e-funnel trap: Automatic monitoring of lepidoptera; a case study of tomato leaf miner. Computers and Electronics in Agriculture. 2021;**185**: 106154

[33] Johari SNAM, Khairunniza-Bejo S, Shariff ARM, Husin NA, Masri MMM, Kamarudin N. Detection of bagworm infestation area in oil palm plantation based on UAV remote sensing using machine learning approach. Agriculture. 2023;**13**(10):1886

[34] Bai L, Huang X, Dashzebeg G, Ariunaa M, Yin S, Bao Y, et al. Potential of unmanned aerial vehicle red–green–blue images for detecting needle pests: A case study with *Erannis jacobsoni* Djak (Lepidoptera, Geometridae). Insects. 2024;**15**(3):172

[35] Mohd Johari SNA, Khairunniza-Bejo S, Mohamed Shariff AR, Husin NA, Mohd Masri MM, Kamarudin N. Effect of datasets size on the machine learning performance of the bagworm, Metisa plana (Walker) infestation using UAV remote sensing. Journal of Plant Diseases and Protection. 2025;**132**(1):1-17

[36] dos Santos LM, Ferraz GAES, Bento NL, Marin DB, Rossi G, Bambi G, et al. Use of images obtained by remotely piloted aircraft and random forest for the detection of leaf miner (*Leucoptera coffeella*) in newly planted coffee trees. Remote Sensing. 2024;**16**(4):728

[37] Dong X, Sun X, Jia X, Xi Z, Gao L, Zhang B. Remote sensing image super-resolution using novel dense-sampling networks. IEEE Transactions on Geoscience and Remote Sensing. 2020; **59**(2):1618-1633

[38] Schütz S, Schöning MJ, Schroth P, Malkoc Ü, Weissbecker B, Kordos P, et al. An insect-based BioFET as a bioelectronic nose. Sensors and Actuators B: Chemical. 2000;**65**(1–3): 291-295

[39] Alberti S, Stasolla G, Mazzola S, Casacci LP, Barbero F. Bioacoustic IoT sensors as next-generation tools for monitoring: Counting flying insects through buzz. Insects. 2023;**14**(12):924

[40] Salehi SS, Rajabpour A, Rasekh A, Farkhari M. Repellency and some biological effects of different ultrasonic waves on Mediterranean flour moth, *Ephestia kuehniella* (Zeller) (Lepidoptera: Pyralidae). Journal of Stored Products Research. 2016;**69**:14-21

[41] Agah-Manesh H, Rajabpour A, Yarahmadi F, Farsi A. Potential of ultrasound to control *Sesamia cretica* (Lepidoptera: Noctuidae). Environmental Entomology. 2021;**50**(6): 1393-1399